Ernest Anye Fongwa

Business Modelling Using Petri Nets Approach

Ernest Anye Fongwa

Business Modelling Using Petri Nets Approach

Preserving ecosystem services by Community-based financial participation

Südwestdeutscher Verlag für Hochschulschriften

Impressum/Imprint (nur für Deutschland/only for Germany)
Bibliografische Information der Deutschen Nationalbibliothek: Die Deutsche Nationalbibliothek verzeichnet diese Publikation in der Deutschen Nationalbibliografie; detaillierte bibliografische Daten sind im Internet über http://dnb.d-nb.de abrufbar.
Alle in diesem Buch genannten Marken und Produktnamen unterliegen warenzeichen-, marken- oder patentrechtlichem Schutz bzw. sind Warenzeichen oder eingetragene Warenzeichen der jeweiligen Inhaber. Die Wiedergabe von Marken, Produktnamen, Gebrauchsnamen, Handelsnamen, Warenbezeichnungen u.s.w. in diesem Werk berechtigt auch ohne besondere Kennzeichnung nicht zu der Annahme, dass solche Namen im Sinne der Warenzeichen- und Markenschutzgesetzgebung als frei zu betrachten wären und daher von jedermann benutzt werden dürften.

Verlag: Südwestdeutscher Verlag für Hochschulschriften GmbH & Co. KG
Heinrich-Böcking-Str. 6-8, 66121 Saarbrücken, Deutschland
Telefon +49 681 37 20 271-1, Telefax +49 681 37 20 271-0
Email: info@svh-verlag.de

Approved by: Cottbus, BTU, Dissertation, 2011

Herstellung in Deutschland:
Schaltungsdienst Lange o.H.G., Berlin
Books on Demand GmbH, Norderstedt
Reha GmbH, Saarbrücken
Amazon Distribution GmbH, Leipzig
ISBN: 978-3-8381-3143-6

Imprint (only for USA, GB)
Bibliographic information published by the Deutsche Nationalbibliothek: The Deutsche Nationalbibliothek lists this publication in the Deutsche Nationalbibliografie; detailed bibliographic data are available in the Internet at http://dnb.d-nb.de.
Any brand names and product names mentioned in this book are subject to trademark, brand or patent protection and are trademarks or registered trademarks of their respective holders. The use of brand names, product names, common names, trade names, product descriptions etc. even without a particular marking in this works is in no way to be construed to mean that such names may be regarded as unrestricted in respect of trademark and brand protection legislation and could thus be used by anyone.

Publisher: Südwestdeutscher Verlag für Hochschulschriften GmbH & Co. KG
Heinrich-Böcking-Str. 6-8, 66121 Saarbrücken, Germany
Phone +49 681 37 20 271-1, Fax +49 681 37 20 271-0
Email: info@svh-verlag.de

Printed in the U.S.A.
Printed in the U.K. by (see last page)
ISBN: 978-3-8381-3143-6

Copyright © 2012 by the author and Südwestdeutscher Verlag für Hochschulschriften GmbH & Co. KG and licensors
All rights reserved. Saarbrücken 2012

Brandenburg
University of Technology
Cottbus

Faculty of Environmental Sciences and Process Engineering
International Course of Environmental and Resource Management

PhD Thesis

A Business Modelling Approach Using Petri Nets to Preserve Ecosystem Services by Community-based Financial Participation

A thesis approved by the Faculty of Environmental Sciences and Process Engineering at the Brandenburg University of Technology in Cottbus, Germany, in partial fulfilment of the requirements for the award of the academic degree of Doctor of Philosophy (Ph.D.) in Environmental Sciences.

by

M. Sc. Ernest Anye Fongwa

Place of birth: Mankon, Bamenda- Cameroon

Supervisors:

Prof. Dr. Albrecht Gnauck
Dept. of Ecosystems and Environmental Informatics
Brandenburg University of Technology at Cottbus, Germany

Prof. Dr. Felix Müller
Ecology Centre
Christian-Albrechts-University of Kiel, Germany

Day of the oral examination: 30.11.2011

Brandenburgische
Technische Universität
Cottbus

Fakultät Umweltwissenschaften und Verfahrenstechnik
Studiengang Environmental and Resource Management

Doktorarbeit

Modellierung von Geschäftsprozessen mit Petri-Netzen zur Erhaltung von Ökosystemdienstleistungen auf der Grundlage einer finanziellen Beteiligung von Gemeinden

Von der Fakultät für Umweltwissenshaften und Verfahrenstechnik der Brandenburgischen Technischen Universität Cottbus zur Erlangung des akademischen Grades eines Doktors der Philosophie (Ph.D.) in Umweltwissenchaften genehmigte Dissertation

vorgelegt von

M. Sc. Ernest Anye Fongwa

Geboren in: Mankon, Bamenda- Kamerun

Betreuer:

Prof. Dr. Albrecht Gnauck
Lehrstuhl Ökosysteme und Umweltinformatik
Brandenburgische Technische Universität Cottbus

Prof. Dr. Felix Müller
Ökologie Zentrum
Christian-Abrechts Universität zu Kiel

Tag der mündlichen Prüfung: 30.11.2011

Declaration

I, Ernest Anye Fongwa, hereby declare that this PhD thesis is my own work, which is completed under the supervision of Prof. Dr. Albrecht Gnauck and Prof. Dr. Felix Müller, and that, to the best of my knowledge and belief, it contains no material previously published or written by another person nor materials which to a substantial extend has been accepted for the award of any other degree or diploma of other universities or institutes of higher learning, except where due acknowledgement has been made in the text.

Date: ………………….. Signature: ………………….

Acknowledgements

I am very grateful for the opportunity that Prof. Dr. Albrecht Gnauck gave to me to write this Ph.D. thesis under his supervision and formulating the research topic as well. This came as a result of a lot of discussions with him on his research recommendations on the use of Petri nets to model ecological processes and the recommendations of my master thesis on financial participation for investing in business development to protect the environment. His supports throughout the research for this thesis are enormous and I will like to highly acknowledge them. I appreciate his continuous comments and suggestions that have helped to improve the thesis. Also, for making it possible the presentation of the thesis results to other scientists for their feedback for improvement. In addition, he made it possible the participation in many conferences and workshops, which empowered my knowledge on the subject matter that were essential for improvement of the thesis. He provided me with teaching and research assistantships within his Department of Ecosystems and Environmental Informatics, which have also helped to partially finance this thesis.

Thanks to Prof. Dr. Felix Müller for accepting to be the second supervisor of this thesis. I highly also appreciate his supports in this thesis that were valuable for improving it. His suggestion on the inclusion of landscape cover data using GIS representation and also giving me a working week together with his staffs on brainstorming of the possibility of mapping ecosystem services within a landscape scale is highly acknowledged. In addition, I will like to thank him for providing a good review on the specific papers that were written within the framework of the research for this thesis and also for reviewing the thesis as well. Thanks also to Dr. Benjamin Burkhard and Mrs. Franziska Kroll, the work we did in Kiel was also helpful for this thesis.

I will like to acknowledge the support of Mr. Michael Petschick for the constant discussions on possibilities of environmental protection at the Spreewald Biosphere Reserve and his invitation to have the possibility to participate in discussion meetings on sustainable development in Spreewald like in the certification scheme, Spreewald Union (Verein) board meeting and others. His supports in the research for this thesis are also enormous. Thanks also go to Prof. Dr. Monika Heiner for supporting this research with a free version of Snoopy (Snoopy is developed under her Department of Data Structure and Software Dependability, BTU Cottbus), which is the main tool for the methodological framework of this thesis. I will equally like to thank Mr. Fei Liu for given me technical assistant on Snoopy and also for the time we had together to brain stream on coloured Petri nets.

In addition, I will like to thank Mrs. Gabriele Richter for her continuous words of encouragement, especially during times of financial difficulties within this research. Her support in corresponding with contacts to institutions and conferences that I visited within the framework of this research were also very helpful. I thank her again for all the nice things she brought to me from her garden. The technical support of Dipl.-Ing. Mirko Filetti is also acknowledged, and his positive comments and suggestions on the research during the colloquium presentations of the research results at the Department of Ecosystems and Environmental Informatics at BTU Cottbus were very helpful for improving the thesis as well. The support of Dr. Bernhard Luther for the brain streaming on the possibilities of normalising data collected by observation is highly appropriated.

I will like to appropriate the support of Prof. Dr. Gerhard Wiegleb for nominating me to teach SPSS for data analysis as part of the course Introduction to Environmental and Resource Management at BTU Cottbus for Master students for 3 winter semesters. It was also helpful for partially financing my PhD research. The support of the Department of Ecosystem and Environmental Informatics at BTU Cottbus for providing me with a working space, materials and equipments for this research is also acknowledged. I appropriate the support of the International Academic Office at BTU Cottbus for constantly nominating my application for teaching and research assistantship within the framework of the financial support programme STIBET advertised by the DAAD. In addition for nominating my application to participate in the internationalisation of research under the financial support programme of the State of Brandenburg Ministry of Science, Research and Culture (MWFK). Thanks also goes to the Spreewald Union (Spreewaldverein) for giving me the possibility to participate in their board meeting and also giving me financial support to enable me complete my thesis. Thanks also to those who provide me with information during my field surveys, interviews and focus group discussion within the UNESCO biosphere reserve Spreewald and including the various administrative offices in Spreewald region and State of Brandenburg, their supports are also acknowledged.

Furthermore, I also acknowledge all those who supported the research for thesis with comments and suggests for improvement during the Ph.D. seminar presentations, especially apl. Prof. Dr. Manfred Wanner (thanks for the great encouragement), as well as during conferences and workshops were the research results were presented. Finally, I will like to thank my family, wife, relatives and friends for their support, encouragement and patient throughout the period of my studies and the research for this thesis.

Abstract

Nature protection has been a challenge for a long time due to continuous activities that reduced ecological processes and functions of the natural environment. These ecological processes and function provide huge benefits to humans and their communities known as Ecosystem Services (ES). ES are the benefits people derive from natural ecosystems. Many policy measures such as legal and voluntary protections have been used to preserve ES. But continuous increasing threats on the natural environment show that additional policy instruments are needed to achieve targets for environmental protection and sustainability. There are increasing land degradation, desertification, flooding, water shortage, climate change, which is mainly due to the sinks in ES. The problem is that ES are often considered as free goods making it different for their preservation. Even though tax policy for payment for ES has been used to encourage their preservation, they are reducing in supply. This is due to social dilemmas like differences in preferences, free riders and "tragedy of the Common" problems that are usually the case in resource distribution. Therefore tax systems may not lead to substantial preservation of ES. This has been argued on using utility theories and introduced market-base instruments for preserving ES as one way of achieving policy targets for nature protection. But for this Market for Ecosystem Services (MES) to grow, incentive and motivation schemes are needed to encourage multi-actors interaction in them. This has led to the establishment of a Community-based Financial Participation (CFP) framework as incentive and motivation schemes for business development for preserving ES.

This thesis uses socio-economic and ecological concepts to argue for MES and developed a business modelling framework with Petri net. This involves multi-actors' interaction for the demand and supply for ES (multi-agent interaction) that are modelled for estimating their flows over a landscape for supporting policy targets based on market structures. It also modelled future scenario for achieving environmental balancing for preserving ES and provide a management framework for MES. The management scheme comprises of an estimating model for ES, and models for management scheme and CFP. This provides the basis of using Petri net modelling framework for CFP to foster the growth of MES as options for preserving ES.

Petri nets modelling techniques is used in simulating multi-agent interaction for the demand and supply of ES that are couple with management and CFP models to form a unified system approach. Petri nets have been used in many disciplines, but this is the first attempt in its use for modelling ES. CFP has also been developed with the framework of the research for this thesis. UNESCO Biosphere Reserve Spreewald (BRSW) in the Lusatian region of south Brandenburg in Germany is taken as an experimental area for testing the concepts and models developed within the thesis research. The results of this thesis show that MES is a potential for achieving targets for preserving ES, which can contribute to nature protection. This has been justified through data sampling, analytical schemes and management support systems that have been provided in the business modelling and simulation framework in this thesis.

Keywords: Ecosystem services, business modelling, Petri nets, community-based financial participation, UNESCO Biosphere Reserve Spreewald

Table of Contents

Declaration		i
Acknowledgement		ii
Abstract		iv
Table of Contents		v
List of Figures		vii
List of Tables		ix
Glossary of Symbols and Abbreviations		x
1.	**Introduction**	**1**
2.	**Theoretical Background**	**10**
2.1	Ecosystem Services	10
2.2	Payments for Valuation of Ecosystem Services	16
2.3	Business Development for Preserving Ecosystem Services	18
2.3.1	Market Structures for Preserving Ecosystem Services	19
2.3.2	Market for Carbon Sequestration	22
2.3.3	Market for Water	23
2.3.4	Market for Biodiversity	25
2.4	Trading Platforms to Establishing Markets for Preserving Ecosystem Services	28
2.5	Community-based Financial Participation	29
3.	**Methodology**	**33**
3.1	Petri Net Modelling Framework	33
3.1.1	Conceptual Model Building with Petri Nets	36
3.1.2	Petri Net Relationships	40
3.1.3	Formal Introduction of Net Relationships	42

3.1.4	Execution and Implementation of Petri Nets Using Snoopy Software	46
3.1.5	Verification and Validation of Properties of the Petri Nets	54
3.2	Data Sampling Strategy	55
3.2.1	Preparatory Set Up for Data Collection	56
3.2.2	Data Aggregation Procedures	57
3.2.3	Quality Assurance of Data	58
3.3	Parameters for Estimating the Values of Ecosystem Services	59
4.	**Model Application to UNESCO Biosphere Reserve Spreewald**	**61**
4.1	Landscape Identification	63
4.2	Data Collection System and Aggregation for Estimating Ecosystem Services	67
4.3	Results of Estimates of Ecosystem Services	69
4.4	Modelling and Simulation Results for Managing Ecosystem Services	74
4.5	Modelling and Simulation Results for Business Development for Preserving Ecosystem Services	81
4.6	Discussions of Results	85
5.	**Conclusions**	**87**
	References	**90**

List of Figures

Figure 1:	Market Framework	20
Figure 2:	Market Framework with a Management Company	21
Figure 3:	Market Framework with Multiple Participants	22
Figure 4:	Community-based Financial Participation	31
Figure 5:	Concept of Business Development for Preserving Ecosystem Services	32
Figure 6:	Four Main Primary Elements of Petri Nets	34
Figure 7:	Petri Net of an Ecosystem Process for Algae Growth	35
Figure 8:	Petri Net Model for Flow of Ecosystem Services	36
Figure 9:	An Extension of Petri Net Model with Multi-agents	37
Figure 10:	Petri Net of Management Scheme for Market for Ecosystem Services	39
Figure 11:	Petri Net of Community-based Financial Participation for Market for Ecosystem Services	40
Figure 12:	Defined Petri Net for Modelling Multi-agent Activities for Estimating Ecosystem Services	49
Figure 13:	A Defined Petri Net for Modelling Management Scheme for Market for Ecosystem Services	50
Figure 14:	Defined Petri Net for Modelling Community-based Financial Participation for Developing Market for Ecosystem Services	51
Figure 15:	Defined Unified Petri Net for Business Modelling of Market for Preserving Ecosystem Services	52
Figure 16:	Map of UNESCO Biosphere Reserve Spreewald with Zones of Protection	61
Figure 17:	Land Use at the UNESCO Biosphere Reserve Spreewald from the Years 1991- 2010	64
Figure 18:	Water System of UNESCO Biosphere Reserve Spreewald	65

Figure 19:	Estimate of Ecosystem Services in UNESCO Biosphere Reserve Spreewald	70
Figure 20:	Potential Demand and Supply of Ecosystem Services for Carbon Sequestration	71
Figure 21:	Potential Demand and Supply of Ecosystem Services for Biodiversity	72
Figure 22:	Potential Demand and Supply of Ecosystem Services for Water Resource	73
Figure 23:	Trend of Ecosystem Services at the UNESCO Biosphere Reserve Spreewald	77
Figure 24:	Modelled Trend of Ecosystem Services by slightly Increased in Supply	78
Figure 25:	Modelled Trend of Ecosystem Services by Higher Supply Level	79
Figure 26:	Trend of Ecosystem Services by Studying Multi-actors actions	80
Figure 27:	Simulation Results for Managing Market for Preserving Ecosystem Services at the UNESCO Biosphere Reserve Spreewald	84

List of Tables

Table 1:	Classification of the Ecosystem Services of Wetland Catchment Areas	12
Table 2:	Categories and Components of Agro-forest Ecosystem Services	13
Table 3:	Indicator System for Agro-forest Ecosystem Services	15
Table 4:	Types of Petri and their Characteristics	34
Table 5:	Summary of Firing Sequence and Formal Petri Net Relationships	43
Table 6:	Definition of Colour Sets for Implementing Petri Nets	46
Table 7:	Summary of Verification of Petri Nets Properties	54
Table 8:	Sampling Units at the Administrate Unit of Straupitz	58
Table 9:	Conversion of Ranking Scale	59
Table 10:	Rate of Demand and Supply of Ecosystem Services by Multi-agents	74
Table 11:	Summary of Dataset Encoded in the Places of a Defined Petri net	75
Table 12:	Summary of Simulated Results Generated with Data of ES from UNESCO BiosphereReserve Spreewald	76
Table 13:	Summary of Simulated Results Generated with Data of ES from UNESCO BiosphereReserve Spreewald	83

Glossary of Symbols and Abbreviations

A	-	Definition of an as ecosystem service A
a	-	a defines a colour for tokens of flow of financial benefits
AAUs	-	Assigned Amount Units
A (on net)	-	Colour for tokens of flow of financial benefits of a Petri net
A_i	-	An incident matrix
ActI	-	Activity impacting ecosystem services
Adj.	-	Adjustment of a quantity or value (Error allocation)
ArcGIS	-	Geographical Information Software
B	-	Definition of an ecosystem service as B
b	-	b defines a colour for tokens of flow of financial contributions
BOD	-	Biological oxygen demand
BNF	-	Backus normal form
BR	-	Biosphere reserve
BRS	-	Biosphere reserve Spreewald
C	-	Consumption
C1	-	Defined colour of a certain set in Petri net
C2	-	Defined colour of a certain set in Petri net
C3	-	Defined colour of a certain set in Petri net
CCS	-	Calculus communicating system
Cd	-	Colour domain
CDM	-	Cleaner development mechanism
CER	-	Corporate environmental responsibility
CERC	-	Certified emission reduction credits
CFP	-	Community-based Financial Participation
COD	-	Chemical Oxygen Demand
Comp	-	Component of ecosystems
$Comp_1$	-	Land surface
$Comp_2$	-	Water
$Comp_3$	-	Biodiversity

CSR	-	Corporate social responsibility
D	-	A defined period of time
DD	-	Demand for Ecosystem Services
DES	-	Discrete event simulation
EASAC	-	European Academies Science Advisory Council
ECOS	-	Terrestrial ecosystems
$ECOS_1$	-	Agricultural ecosystems
$ECOS_2$	-	Forestry ecosystems
$ECOS_3$	-	Wetland ecosystems
$ECOS_4$	-	Plants including green plants ecosystems
$ECOS_5$	-	Urban ecosystems
$ECOS_6$	-	Developed ecosystems
$ECOS_7$	-	Mixed ecosystems
$ECOS_8$	-	Other ecosystems
ERU	-	Emission Reduction Unit
ES	-	Ecosystem services
ES_1	-	Provision ecosystem service
ES_2	-	Regulating ecosystem service
ES_3	-	Supporting ecosystem services
ES_4	-	Preserving ecosystem services
ES_5	-	Cultural ecosystem services
ESI	-	Ecosystem Services Index
ESO	-	Employee Share Ownership
ESOP	-	Employee Share Ownership Plan
ETS	-	Emission Trading System
ε	-	Uncertainty state
Fin	-	Defined a marking finance of a Petri net/ colour set in Petri net
FP	-	Financial Participation
FSC	-	Forest Steward Council
GDP	-	Gross Domestic Product
GNP	-	Gross National Product

g(x)	-	Denote an amount of contribution in a CFP
g´(x)	-	Denote an amount of contribution in a CFP
g´´(x)	-	Denote an amount of contribution in a CFP
H	-	A contribution by an individual
IFOAM	-	International Federation of Organic Agricultural Movements
INC	-	Particulate type of legal entity for corporations
IUCN	-	International Union for Conservation of Nature
K	-	Capital contribution
L	-	Landscape type
M	-	Variable of a colour set of financial flow (Fin) in a Petri net
Ma	-	Multi-actor
MAB	-	Man and the Biosphere
MEA	-	Millennium Ecosystem Assessment
MEAs	-	Multilateral Environmental Agreements
MES	-	Market for Ecosystem Services
MS	-	Management strategies/ marking Management Strategy/Colour set in Petri net
MS_1	-	Defined a token for management strategy "Permit"
MS_2	-	Defined a token for management strategy "Certification"
MS_3	-	Defined a token for management strategy "Conservation Credit/Banking"
$MSMS_1$	-	Colour for tokens Permit
$MSMS_2$	-	Colour for tokens Certification
MSMS3	-	Colour for tokens Conservation Credit/Banking
N	-	Defined a Petri net as N
n_1	-	Defined marking in a Petri net at a place MES
n_2	-	Defined marking in a Petri net at the place supply unit for ES
n_3	-	Defined marking in a Petri net at the place demand unit for ES
n_4	-	Defined marking in a Petri net at the stock of ES
NGO	-	Non-governmental organisation
NPV	-	Net Present Value

NTES	-	Non-Tangible ES
P	-	A set of elements
PEFC	-	Program for the Endorsement of Forest Certification
PEPPER	-	Promotion of European Participation in Profit and Enterprise Results
PP	-	Colour set in Petri net
PPP	-	Public Private Partnership
RMU	-	Removal Unit
S	-	Place of a Petri net
s_1	-	Place of a Petri net "MES"
s_2	-	Place of a Petri net "Supply unit for ES"
s_3	-	Place of a Petri net "Demand unit for ES"
s_4	-	Place of a Petri net "Stock of ES"
SAU	-	Service Antagonising Unit
SPU	-	Service Providing Unit
SS	-	Supply of Ecosystem Services
St	-	Stock of ES
SU	-	Service Units
SWOT	-	Strength Weakness Opportunity and Threat
T	-	Transition of a Petri net
t	-	Time period
t_1	-	Transition of a Petri net "Producers",
t_2	-	Transition of a Petri net "Suppliers"
t_3	-	Transition of a Petri net "Migrants",
t_4	-	Transition of a Petri net "Developers",
t_5	-	Transition of a Petri net "Antagonisers"
t_6	-	Transition of a Petri net "Consumers"
t_7	-	Transition of a Petri net "Supply of ES"
t_8	-	Transition of a Petri net "Demand of ES"
TES	-	Tangible ES
TO	-	Trade-off

ToRs	-	Terms of References
u*	-	Utility function that can take any quantity
V	-	A defined capital consumption value with V
VES	-	Value for ES
VU	-	Value of ecosystem services
W	-	Weight of arc of a Petri net
WTA	-	Willingness To Accept
WTP	-	Willingness To Pay
X	-	Defined an ES "X"
x	-	Number of a Variable
Y	-	An actor for a CFP/ A contribution by an individual

Logical symbols

&	-	And
++	-	A summation sign of the Backus Normal Form (BFN)
$\forall$	-	For all
Σ	-	Summation
$\in$	-	Element of
Ω	-	Denoting an allocation (pareto optimal allocation)
Δ	-	Delta denoting change of a variable
$\mathbb{N}$	-	Natural numbers
Б	-	Permutation
op	-	Opposite or conversion relation of a variable
cd	-	Expression of a colour set
β	-	Disturbance
∞	-	Infinity
$\cup$	-	Union of a set

Chemical Elements/ Compounds

CH_4	-	Methane
HFCs	-	Hydrofluorocarbons
N	-	Nitrogen
O_2	-	Oxygen (Gaseous)
P	-	Phosphorus
PFCs	-	Perfluorocarbons
SF_6	-	Sulphur hexafluoride

1. Introduction

The human society has been continuously living with uncertainty on how their environment will look like in the future due to increase in activities that reduce the supply of Ecosystem Services (ES). Despite increasing environmental studies, campaigns, literatures, workshops, conferences, policies and programmes to create awareness on environmental concerns, ES and its flows that maintain and balance environmental processes is still poorly understood in most areas in the world. That is why they are reducing in supply, even though they are being balance or in plus in some regions in the world, but globally they are reducing in supply (MEA, 2005; Bauer and Stringer, 2009; World Resources Institute, 2008; FAO, 2005 and 2007; Wohlmuth, 2007). Threats continue to impose on the reduction of ES, especially as world population is growing. The increase in world population also means additional pressure on ES due to increase in its demand and the misconception that they are free and infinitely available. Such misconception has led to the abuse on the exploitation of natural resources, leading to destruction of ecosystems and reduction of ES in most regions in the world. Water bodies are being over-fished, pests and diseases are extending beyond human expectations, and deforestation is limiting flood control around human settlement leading to loss of life and properties (MEA, 2005; World Resources Institute, 2008).

In addition, industrial and human activities are polluting water bodies, soil and atmosphere causing increasing environmental problems such as greenhouse gasses leading to climate change. Many integrated studies at the traditional, local, regional and global scales present huge reports of mismanagement of these services (MEA, 2005; World Resources Institute, 2008; MEAs, 2005; World Agro-forestry Center, 2007). In some of the studies, out of 24 ecosystems assessed, only 4 have shown improvement over the past 50 years, 15 are in serious decline and five are almost at a balance (MEAs, 2005). Many analysis and reports from international organizations like the United Nations approximate that about 40 to 50% of the Earth's ice-free land surface has been highly transformed or degraded by anthropogenic activities. About 66% of marine fisheries are overexploited or either at their limit, atmospheric CO_2 has increased more than 30% since the advent of industrialisation, and nearly 25% of the earth's birds and animals species have been extinct in the last 2000 years (Vitousek, 1997; Bauer and Stringer, 2009; World Resources Institute, 2008; FAO, 2005 and 2007; WEO, 2008; World Bank, 2008; EIA, 2008; ESRL, 2008). The MEA (2005) found that 60% of global ES are being degraded or used unsustainably. This is mainly due to consequences of the utilisation of natural resources to satisfy growing consumption of goods and services particularly in the private sector. Therefore, the reduction of ES can be attributed to the action of man and natural disasters on the environment. Agro-forest ecosystems occupy the largest part of land use at the landscape scale. Many studies and results show that they are one of the highest ecosystem types undergoing land degradation due to the constant conversion of forest land for intensification of agricultural activities (Kleijn and Sutherland, 2003; EASAC, 2005). This is not the only case of land degradation. Therefore one can argue that the degree of utilisation of natural resources that support ES is higher than the degree of balance in their supply to protect nature, which can be justified from these reports. Ranaganathan and Irwin (2007) also argued that it is difficult for any development or providing project investment in a community that do not depend upon and affect ES (directly or indirectly). If ES are allowed to decline, so too are the benefits they provide to communities.

However, preserving and conserving ES require that appropriate values be attached to them (Slootweg and Beukering, 2008; Farber et al., 2002; Daily and Ellison 2002), although it is considered that nature offers them free of charge. But the health of environment depends on their preservation (MEA, 2005), which continuous influence of anthropogenic activities on them at landscapes are leading to increasing land degradation. For instance, the intensification of agriculture with inappropriate management measures and control of community waste system would be drive to land degradation (Chamberlain et al., 2000; Kleijn and Sutherland, 2003; Mclaughin and Mineau, 1995; Slokoe and Teague, 1995; Fongwa and Gnauck, 2010). Most governments and international organisations think the solution to such problems of land degradation are protected areas leading to many protected areas around the world (Natura, 2000; Hansen and Defries, 2007). This has also given rise to many biosphere reserves as UNESCO programme of Man and the Biosphere (MAB) to give value to biodiversity and conserve landscapes. A Biosphere Reserve (BR) concept under UNESCO consists of three main levels of restricted areas:

- Core zone, where the most delicate flora and fauna thrive in normally disturbed ecosystems
- Buffer zone, where economic activities can take place as well as environmental education, training and recreation, and
- Transitional areas are where people usually live within a BR and the aim is to protect and ensure the rational use of natural resources.

The idea of BR is in line with the enforcement of the local agenda 21 defined under Rio conference in 1992 of the UN as efforts to support poverty reduction and sustainable development (see http://www.un.org/esa/dsd/agenda21/). Sustainable development requires the balancing of ecological, economic and socio-cultural elements within the present needs, while thinking about the future generation (WCED, 1987; Rao, 2000). This means, if an area is already degraded, then keeping it to only undergo natural dynamics may be a challenge. This type of area would lack the regeneration potential to protect the environment due to a sink in ES. Therefore, protecting it is not in line within sustainable development, but rehabilitation schemes before protection should be the way further. This is usually the case with BRs, where the core protected zone are typical degraded areas. Protecting an area with imbalance in environmental function and processes makes it vulnerable to natural and anthropogenic conditions, especially when they occurred.

It is ethically right that everybody have the right to live and consume natural resources, and it is equally ethnical correct that the environment and ecosystems have to be preserved, not to damage its functionality and services that support life (ES). But poverty in many societies around the world and industrialisation have led to increasing mal-practices for natural resource exploitation that have a negative impact on the environment and ecosystems (World Agro-forestry Center, 2007; Fongwa and Gnauck, 2010). So one can argue that for sustaining ES is a challenge for the future, which can be combat by involving an increasing number of multi-actors in an ecosystem community for their development. This requires identifying appropriate motivation and incentive schemes, and control systems to make the preservation of ES beneficial to all actors in an ecosystem community, which is still challenging. Schemes that provide cooperation within stakeholders in communities (due to multi-actors' behaviour in ecosystems or differences in preference) can allow for payment and maintenance of damages caused on ecosystems, which can be away forward in preserving ES.

Economists and ecologists have begun to describe what potential value can be attached to a particular ES (Slootweg and Beukering, 2008; Freeman, 1993; Wätzold, 2006; Fongwa and Gnauck, 2009a). They employ political and economic arguments to argue for the preservation of ES. This is because payment systems for ES can be innovative instruments for natural resource management, which can be of significant interest to many stakeholders in an ecosystem community. On the other hand willingness to pay (demand side) is one of the biggest challenge for such systems that is present in both the demand and supply sides. This is mainly due to the normative view or argument (free rider effect) for utilisation of public goods and services. The normative behaviour or free rider effects for participation have proven that citizens can only participate in environmental protection and reduction of ecosystem damages, only if they have a stake in the benefits that arise from ES. This is especially monetary benefits, not the functional benefits. This argument has lead to the consideration of market-based instruments for preservation of ES. Those instruments that can generate financial resource, divert funds to environmentally friendly technologies, generate incentives for investment, and increase the involvement of the private sector in environmental protection (UNEP, 2005). Therefore recognising the need for market-based instruments for ES can play a significant role in the implementation of Multilateral Environmental Agreements (MEAs) and spearheading sustainable development and poverty reduction, especially in many developing regions.

Payment and trading of ES are emerging as world-wide small-scale solutions, where one can acquire credits for activities such as sponsoring the protection of carbon sequestration sources or the restoration of ES providers (Daily, 2006). These transactions even though are strategic for preserving ES, they lack sustainability frameworks. Especially when employing networks that permit the estimate of services across landscapes, and also protect them through a diversification of investment. These efforts can be seen as good schemes to preserve ES, but include inconsistent compensation for services or resources sacrificed elsewhere and misconceived warrants for their irresponsible use (Daily, 2006). They have to be organised to fall within sustainability frameworks by defining criteria for estimating them. That is development of strategies that will last for a longer time, feasible and meet the needs of local ownership (EC, 2004). This requires that they should be formulated to capture measures of balancing environmental processes and functions within landscapes and beyond. Research approaches developed in the context of multi-functionality of ES can contribute to the understanding that for sustainable development to be achieved, communities/businesses need to integrate ES in their activities. But the challenge still remains on the determination of tools of integrating communities and business developments within mechanisms for preserving ES. If, it is possible to create and establish an inventory of ES to determine and outline drives that have an impact on ecosystems based on multi-agent interactions. Then this can provide grounds for the development of simple concepts for integrating ES with economic activities in an ecosystem community and the use of analytical tools to understand their changes for policy measures or management strategies.

The integration of ES into business developmental processes can also provide grounds for businesses in acknowledging the aspect of Corporate Social Responsibility (CSR) and Corporate Environmental Responsibility (CER) as a result of its existence in its corporate community. Thereby increasing the goodwill of businesses and equally providing their customers with confidence that they can rely on them, which can even extend beyond their community. This idea is not seen by many businesses as an opportunity because ES are mostly intangibles. It is simple to put a price on say timber harvested from the forest or copper mined from the ground, but difficult to get the economic value on less tangible ES, such as water purification and flood control. The economic value of ES is never zero and can be very large. Several international and conservation bodies, such as the IUCN (International Union for Conservation of Nature) and UNEP are advocating the use of markets and payments for ES. This is to ensure that beneficiaries pay for services and their providers being reimbursed, thereby creating incentives for continuous service provision and the ecosystem protection. The attempt is to establish an individual WTP for ES or avoid its degradation or "Willingness To Accept (WTA)" - compensation for the degradation of ES. The characteristics of various goods and services affect the ease with which market-based tools can elicit their value marketed.

Market goods are most often excludable (a legal concept that allows an owner to prevent another person from using the assets) and rival (where consumption or use reduces the amount available for other people), whereas most ES are non excludable, and less rival (Daly and Farley, 2004). On the other hand, conditions that satisfy market efficiency do not include environmental sustainability or socially just distribution (Daly and Farley, 2004). But excludability focuses on direct benefits, which can obscure the important benefits from distributive justice. To some extend, social agreements can encourage excludability or rivalries or create a proxy (consider "carbon credit scheme") to make ES marketable. This means that market network systems such as tradable credit schemes can make the market for ES to grow big. This would depend on regulations (laws and standards), market incentives, information (like certification) and institutional flexibility can influence the success of attempts to bring market attribute for ES. Therefore business development for ES depends on environmental structures that enable frameworks for preserving them. That are laws, regulations, taxes, subsidies, social norms and voluntary agreements within which businesses and it community operate. For example, for business to value ES, it must ultimately become more profitable to preserve them than ignore or destroy them. In many countries such as Brazil, Australia and Panama with emerging Market for Ecosystem Services (MES), significant reforms have given enabling environment for requirements for business development for ES to grow (Daily and Ellison, 2002). This is particularly where the role of conservation is not limited only to law, or where policy incentives such as preservation subsidies are causing continuous harm to ES maintenance and development.

Nevertheless, there has been growing recognition of commercial strategy for ES within environmentalists and the business communities. But the question is whether businesses and investors really considered ES in their business decisions. Many companies still have trouble seeing the bottom-line relationship between their business and ES. A big challenge for business development for preserving ES is the lack of the understanding between the world of business and nature conservation. For instance, scientists working on natural science often lack the financial acumen (skill in making correct decision and judgement in a particular subject) and consumer orientation of the private sector; conservationists typically lack business planning and management skills. Therefore, business for ES may be viewed as a risk or liability rather than a potential profit centre. This perception is beginning to change as increasing number of companies see a business advantage in developing processes to integrate ES into their operations, as well as seeking market-based solutions and opportunities. The problem is that most business people lack the understanding of how their companies operations affects and are affected by ES, or how to integrate them in their operation. Therefore bringing business development for ES into light, market forces can be used to make them a valuable proposition for investment and commercial-basis. This is because growing markets for ES creates ample opportunities for local and rural entrepreneurship, meaning that business models for preserving ES may reduce rural poverty, provide employment and generate profit.

It has an additional benefit that it can also easily stimulates a flow of funds from relatively wealthy urban centres to the countryside, as well as from industrialised to developing nations. One can argue that if businesses are not integrated into the development and preservation of ES, then businesses can develop other type of structure to run away from environmental liabilities. This would lead to a situation that costs of damages to ecosystems are left to the society (externalities). The reason for business investment varies. These ranges from those companies such as ecotourism initiatives, whose businesses directly depends on healthy ecosystems to corporate level of large scale infrastructure projects or extractive industries that damage or operate in increasing cost for the environment. This is due to policies and trends that create new business opportunities. Therefore, there is a need for the development of a business model for preserving ES as a tool that can clarify the situation and bridge the gaps between planning, management and performance assessment. Such a model should take into consideration social justice to provide insurance for stakeholders, so that they must not fear exploitation from other individuals in the demand and supply of ES. Social justice is a general conception of equalitarianism to compensate persons for inequalities of circumstances that they are not responsible (Roemer, 1996). But this should not be as a result of inequalities from the exercise of preferences that they are responsible. Therefore, a strategy of Pareto optimality has to be provided such as a Community-based Financial Participation (CFP) for the development of businesses for preserving ES. Pareto optimum strategy for ES requires that stakeholders of a community come together to develop and maintain services that they benefit from them based on an optimal level of participation that is acceptable to all of them.

CFP is an incentive and motivation scheme for encouraging increase community-based participation in financial portfolio for investment in business development (Fongwa and Gnauck, 2009b and 2009c; Fongwa et al., 2009 and 2010a). It can lead to a cooperation of all stakeholders in the community to be involved in the preservation of ES (supply side) and benefit in their supply (demand side). This together can lead to the development of them (market side) and benefits (profitability). It can also strongly compromise conflicts that might arise within different interest groups involved in the demand and supply of ES due to potential concessions. The linkages between communities and business developments within mechanisms for preserving ES can also provide grounds for the development of simple concepts for integration and the use of analytical tools to understand their changes for policy measures or management strategies. A modelling framework can contribute in this respect. Modelling is not particular to any discipline; it consists of describing, simplifying, validating and simulating a system to explore their relationship for particular objectives (Zeigler, 1976). It may be to understand the behaviour of the relationship for decision support system or optimisation of variables. Systems are modelled in terms of their state either through each point in time by describing entities that represent the system resources, activities and events (multi-agents) causing the system state to change over time or continuously (Banks et al., 1999; Koji et al., 2007, Fongwa et al., 2010b; Ajmone et al., 1995). The idea is to estimate and provide market-based strategy for preserving ES at the landscape scale through the introduction of management and incentives schemes. Landscapes contain different types of ecosystems and ecological components as well as natural actors with different behaviours. Their complex interactions determine the states of ES (Grime, 2007). The problems of preserving ES can be classified by three aspects: Ecological, economical and social aspects. A more detailed explanation is given in section 2. To manage ES at the landscape scale by economic concepts lead to the following research questions:

- How can estimates of ES at the landscape scale be described by mathematical expressions for encouraging business development for their preservation?

- How can business development for preserving ES support their management at a landscape scale?
- How can CFP be an incentive and motivation scheme to involve public participation in business development for preserving ES?
- How can CFP procedure be modelled to support market-based strategies or policy measures to preserve ES?
- How can a modelling framework be used as a unified system for management support on preserving ES at the landscape scale?

The main aim of this research is to contribute towards nature protection by providing an understanding and description of the linkages that exist between multi-agent interactions for ES, their preservation, payment and markets for them. The linkages are essential for establishing business development strategies for preserving ES using CFP. CFP can provide funding for establishing MES and can motivate multi-stakeholders in preserving ES. The concept of CFP has not existed before. It has been developed within the framework of the research for this thesis. A business modelling approach can make it possible an understanding of the interactions of multi-agents in the demand and supply of ES for establishment of markets to preserve them. That is, the activities that decreases or increases ES can be balance at the landscape based on market conditions with well structured trading schemes and financial incentives as unified business modelling system. The objectives are to encourage business development for preserving ES and provide a computer based model with Petri Nets for their strategic management at the landscape scale. Petri Nets have been used to model and simulate many systems. They have been used in transport, supply chain system, water ecology, molecular biology and many more, but not for ES. This is the first attempt to model ES with Petri Nets in the research under which this thesis is based. It is used because it gives the possibility to compose large and complex systems in to smaller units (parallel composition) for studying them separately, then unifying them as aggregates within the same modelling framework (divide and conquered strategy). It use in modelling ES at the landscape gives the possibility of composing different activities that impact landscape functions and processes as a group of separate units. Then they are brought together to show linkages to ecosystem components that provide ES.

This type of modelling approach for ES makes it possible the presentation of results that can be used by ecosystem developers, researchers, the business community, governmental and consulting agencies, international and funding organisations, and local communities as well. This can encourage their preservation at the landscape. The thesis is structured with the first section giving an introduction of the subject matter that will be examined. It provides a problem analysis that led to the consideration of some research questions to be answered in this thesis with an aim and objectives. The second section provides a theoretical background for understanding concepts and theories that are being examined. It will explain what ES are and provides an indicator system for them. Payment schemes for valuation of ES will also be given that are essential for establishing business discussions for their development, which MES and CFP as incentives for preserving them at the landscape are presented. Section three presents a methodology of Petri Net modelling framework and data sampling strategy. It provides a coloured Petri Net modelling framework by establishing conceptual models with Petri Nets and describing their relationships. This will be essential for their execution and implementation using Snoop software. A data sampling strategy for input variables in the Petri Nets will be also given in this section. It comprises of the data collection and aggregation procedures, and quality assurance for correctness. The section ends up with parameter for estimating values of ES, which will be essential for interpreting the results for decision support system for balancing them on the landscape by business development strategy. The fourth section provides a model application to the UNESCO BRS. It will start by identifying the landscape of the region and then explaining how the data sampling strategy is applied. The modelling framework with Petri Nets will be also applied in this section and provides discusses of the results. The fifth section concludes the thesis. It provides a debriefing on the complexity of ecosystem in providing ES that requires an appropriate data management system, which this thesis will realise and explained how the research questions that were established were answered. In addition, it provides hints for future research in this field and gives an ending remark of the importance of the results of this thesis to nature protection. Finally the thesis ends up with a list of references. The internet publications are separated from the main list of references.

2. Theoretical Background

Many scientists trying to understand how natural processes and functions of the environment are linked to human interactions to foster their preservation have led to the emergence of the topic ES. Therefore, in this chapter the concept of ES will be discussed in relation to the emergence of ecology and economics (Costanza, 1991). It's significant received attention since the appearance of the MEA (2005), which has been attracting the scientific community for its strategic importance for sustainable development and the fight against environmental degradation, desertification and climate change. It is being agreed that ES are critical to sustainable development and that they ought to be preserved, if not enhanced (Vitousek, 1997; Challen, 2000; World Resource Institute, 2008).

ES are today threatened and their importance is undermined. They will be discussed and described in detail in section 2.1. ES offer an opportunity for financial incomes to landscape managers and users. Therefore valuing ES can provide payment from them, which can also encourage business development for their preservation. This can be important for conservation of natural ecosystems and also be an economic incentive. A payment scheme for valuation of ES is discussed in more details in section 2.2. Section 2.3 presents explanations for business development for preserving ES. Market structures and MES are presented, but for business development for them, tradable platforms have to be established that is discussed in section 2.4. Bringing many stakeholders interests together can give the chance for issues and opinions to be raised on potentials business development for preserving ES. It can be an avenue for potential knowledge and information sharing for common benefit based on cooperation for mitigating some activities that are detrimental to preserving ES. But involvement of a broader participation of ecosystems stakeholders through incentive and motivation schemes can potential encourages business development for them, which CFP is an option. CFP is discussed in section 2.5 and an established relationship for it use for developing ES at the landscape is also presented.

2.1 Ecosystem Services

The natural environment provides a multiple of resources and processes that provide benefits to human beings; and these resources and processes are collectively known as ES (Costanza et al., 1997; MEA, 2005; Daily, 1997). They are linked to physiochemical processes of the natural environment (Ranganathan et al., 2008). Therefore, without the ES regulating the global climate and temperatures year round, it is possible that the earth's temperature would fall outside the range that is convenient for human survival; or liquid water would become scarce. That is why there is the constant discussion on climate change. This means staying warm would be a big problem that could cost a lot of money on technologies for heating or cooling, which themselves rely on ES (Farber et al., 2002; Daily and Ellison, 2002). When these ES do not function anymore, it would be for instance possible to have other scenarios in which there would be no air or no water or no food. This is because the species that pollinate our crops do not exist any more. It is only when the role and value of ES is understood that humanity would be making more effort to invest in preserving them within levels requisite for sustaining quality human life (Farber et al., 2002; Daily and Ellison, 2002). Therefore ES are the benefits which people obtain from ecosystems that enhance their socio-economic and cultural well being (Dasgupta, 1996; MEA, 2005; Boyd and Banzhaf, 2007).

ES can be subdivided into five main types with different categories under each of them. They are spread from one ecosystem to another depending on landscape characteristics such as water availability. Their distribution is on a local scale, but their supply and availability depend on the type of ecosystem in a particular catchment area that can then be aggregated from local regional to global scales. A catchment area or drainage basin is an area drained by a stream, river or other small running water bodies (De Barry, 2004). The limit of a given catchment area is its height of land and water systems, which provides a scale of determining the availability of different types and categories of ES. This is based on their physical conditions that can be used to differentiate catchments. Therefore there are different types of catchment areas, but the focus here is on wetland. For instance, the abundance of ES in an agro-forest ecosystem of a wetland catchment area would differ from another agro-forest ecosystem in a dry land catchment or that from a valley would differ from that from a hilly area. Therefore an examination of agro-forest ecosystems to streamline the ES that exist in a particular region also requires the specification of the type of catchment areas. This can provide an understanding on their availability and abundance over particular scales. In table 1 a classification of ES of wetland catchment areas is presented. The services are divided in into types and categories with appropriate examples. The classification of ES into categories and components that rendered their availability based on particular catchment areas can make it possible an inventory system for them. They are many types of ecosystems within a particular wetland catchment area that provide ES, but the focus here is on the agro-forest services. Agro-forest ecosystems are one of the types of ecosystem that occupies the largest land use at the landscape in many parts of the world.

Table 1: Classification of the Ecosystem Services of Wetland Catchment Areas

Types of ES	Category of ES	Selected Examples
Provisioning Services	Food	Crops, livestock, fisheries, aquaculture, wild foods
	Water	Fresh water, rivers, sea and oceans
	Fibre, wood	Timber, cotton, hemp, silk, fruits, nuts
	Energy	Biomass, photosynthesis, solar-rays, oil plants, hydrothermal, geothermal, tidal wave energy, hydro-carbons, fuel wood
	Bio-chemicals	Biomedical plants, bio-remediation compounds, herbs, aromatics, bee wax
Regulating Services	Air quality regulation	Climate and pollutants regulation, filtering of dust particles, odour
	Water quality and quantity regulation	Water purification such as removal of impurities in water like nutrients e.g. exceed phosphate, salinity, acidity (water softening), evapotranspiration, precipitation, water bodies
	Pests and diseases regulation	Regulation against parasite, fungi and bacteria invasion, air and water born diseases, exposure to poisonous substances
	Regulation of soil and erosion	Regulate soil depletion through soil buffering, removal of impurities in soil through soil shrinking, control of wind and water erosions
	Natural hazards regulation	Storm and flood control, control of tectonic movement in soil to reduce earthquakes and volcanic explosion
Supporting Services	Nutrient cycling and soil formation	Nutrient balancing like mineral and other substances through borrowing activities by micro-organisms, soil formation by decomposition of dead plants and animals
	Crop pollination	Pollination by bees, insects and other micro-organisms
	Supporting the earth surface	Life on earth, platform for houses, farming, road and highways, water
Preserving Services	Biodiversity	Habitat for plant and animal species, noise reduction
	Genetic resources against uncertainty (extinction)	Richness/abundance of genetic species, bio-refugia, hybridisation of species
Cultural Services	Spiritual and educational values	Secret places, inspirations, religious grounds, social relations (indigenous culture), sense of place (cultural identity), literature such folk law and stories writing
	Recreational and ecotourism	Aesthetic values, monuments, sanctuaries, natural parks, natural and cultural tourism, heritage sites, sporting sites

In table 2 a categorisation of agro-forest ES and components that determine their availability over a catchment area and activities that have an impact on them is presented (EASAC, 2005; Ngobo, 2004; Banzhaf and Boyd, 2005). The activities lead to their improvement, deficit or balance. This type of categorisation (in table 2) can be used to identify indicators for variables of ES that support inventory systems for estimating their value over specific geographic regions for preservative measures.

Table 2: Categories and Components of Agro-forest Ecosystem Services

Categories of ES	Components of ES	Natural and Human Activities
Food, water, fibre, wood, energy, bio-chemicals	Plant (Tree, grass and crops) and animal species, farmland and forestland, water balance, abandonment of pastoral system, herbs (pharmaceutics), fruits, vegetables	Irrigation, evapotranspiration, resource extraction activities, runoff and drainage, grazing and cultivation, tree removal without replanting, animal breeding, stock feeding, planting and modification of cultivation practices
Air quality regulation, water quality and quantity regulation, pests and diseases regulation, regulation of soil and erosion, natural hazards regulation	Water quality, filtering water, temperature reduction, pest, diseases, carbon sequestration, filtering dust particles air, pollution removal	Pollution by nitrates and pesticides, salinity, acidity and nutrients (P,N and others), production of renewable energy from agro-forestry, greenhouse gas emission and ammonia from agro-forestry, avoidance of emission, odour
Nutrient cycling and soil formation, crop pollination, support the earth surface	Maintenance of soil quality, nutrients balances, bees population, soil loss, nutrient	Extension of bee, organic farming, ecological farming, feed, fertilizer, burning, energy and acquisition of land, weed, animals pests, chemical release on land, areas of risk of soil erosion, cultivation
Biodiversity, genetic resources against uncertainty (extinction)	Storm mitigation, biological refugia, noise reduction, flood control, richness/abundance of species	Deforestation, land fragmentation, conservation banking, hunting
Spiritual and educational values, recreational and ecotourism	Aesthetic, monuments sanctuaries, natural parks, secret and inspirations places, eco-tourism	Tourist activities, cultural activities, transport, birthplace for various traditional performances, cultural heritage

There are many existing literatures with inventories and indicator systems for ES (Ten Brink, 2000; MEA, 2003 and 2005; EEA, 2003 and 2007; Chambers and Lewis, 2001; EASAC, 2005). But their estimate over a particular scale has not yet been clarified, especially their change over time or transition events. Therefore quantifying them to an aggregate for an indicator may also not capture all elements being measured due to mismatches of services over a spatial scale (EEA, 2007). The already existing indicator systems for ES in literatures range from species richness and abundance to pressure state, and adaptive indicators at different scales that can be used to derive measurable quantities for them. The following are parameters for an indicator system that can be used for measuring ES over an ecosystem at a particular landscape scale:

- Species richness based on the number of species availability on ecosystems
- Species abundance based on population size and density on ecosystems
- Pressure state comprising of internal and external activities taking place that influence and impact natural processes on ecosystems, and
- Adaptive or resilience to pressure states based on measures taken by human actions for readdressing negative impacts and their effect to them on ecosystems

Species richness and abundance are aggregated to form a composite indicator for species availability. This means species availability, pressure state and adaptive indicators are the indicator systems that can be formulated for measurability as follows:

- Species availability indicator = Σ species available × pop size/area
- Pressure indicator = $\Sigma\{$Natural and human activities impacting ecosystems ± influence (internal and external effects)$\}$ ± intensity/ area, and
- Adaptive state indicator = $\Sigma\{$Resilience to Pressure state ± influence (internal and external effects)$\}$ ± intensity/ area

The interactions of different activities taking place on ecosystems have a positive or negative impact on indicators. Therefore these activities changing the state of ES have to be measure to determine whether the negatives impacts outweigh the positives impacts. This can be used to justify an environmental degradation hypothesis, such as water shortage, increase in greenhouse gases potentials and others. These indicators are used for estimating ES, which their measurability, aggregation and interpretability is discussed in more details in the next section on methodology under sampling strategy (see section 3.2). In table 3 an indicator system for agro-forest ES is presented. It is based on inventory studies of potential ES in literatures with Term of References (ToRs) to the agro-forestry region at the UNESCO Biosphere Reserve Spreewald (BRS) and synthesis from database with indicator systems (EASAC, 2005; Natura, 2000; OCED, 2001).

Table 3: Indicator System for Agro-forest Ecosystem Services

Categories of ES	Indicators
Provisioning Service:	
Food	Land use, land cover, food supply, food types, ploughing, fertilisers use, land tenure, mixed farming, irrigation, intensive and mechanised agriculture, grazing
Water	Land cover, water flows, fresh water supplies, water consumption, water table, evapotranspiration, lakes, block rivers, water channels, dry streams, swamps, dry landscapes, canalisation
Fibre, wood	Land cover, fibre and tree quantity, types of fibres and trees
Energy	Energy capture (net primary energy production), entropy export/import, energy budget, solar and wind power parks, energy crops
Bio-chemicals	Types and amount of herbs and pharmaceutics, bee wax, bio-chemical and remediation plants
Regulating Services:	
Air quality regulation	Pollutant sequestration capacity by plants and soil, dust removal capacity, reforestation, deforestation and afforestation, dead trees, green plants
Water quality and quantity regulation	Eutrophication, leaching of organic chemicals in water, water retention capacity, runoff, drainage, alkalisation and salinity, algae bloom, floods, flood breaks
Pests and diseases regulation	Availability and types of pest and diseases, use of herbicides and pesticides
Regulation of soil and erosion	Microbial soil respiration, land fragmentation, land bareness, soil compatibility
Natural hazards regulation	Landslides, earthquakes, cyclones, volcanoes, tornadoes, natural dryness, climate, winds
Supporting Services:	
Nutrient cycling and soil formation	Carbon and nitrogen net mineralization, metabolic efficiency, photosynthesis
Crop pollination	Type and availability of bees species, insects and other micro-animal species
Supporting the Earth surface	Land cover, structure of human development, settlement, natural development, canalisation, water reservoirs, storage tanks, densities
Preserving Services:	
Biodiversity	Types and availability of plant and animal species
Genetic resource against uncertainty	Types and availability of genetic species
Cultural Services:	
Spiritual and educational values	Traditions (indigenous culture), religion, sanctuaries, regional produces (cultural identity), learning grounds and potential for stories writing and poets
Recreational and ecotourism	Tourist sites, eco-tourism, availability of tourist services, leisure sites (parks, playing grounds), tourist facilities, diversity of landscape, tourists, aesthetics, tourist establishments

2.2 Payments for Valuation of Ecosystem Services

Many decisions for preservation of ES are taken based on substantial and procedural rationality. This can provide strategies that do not take into consideration those irrational behaviours of stakeholders with adequate intervention mechanisms for preservation of ES that may be detrimental to conservation. This can be acknowledged if one tries to provide answers to the following questions:

- How much could one pay for a sip of water?
- What about a breath of fresh air or handful of fertile soil?
- What about the warm breeze or blue sky worth in monetary terms?

These questions seem difficult, but who will pay for them if should in case they become scarce or would they be priceless? Or would they not become the must valuable and expensive product in the world, which can be answered with economic concepts of scarcity and choice (Barr, 2004; Stiglitz, 2000). Therefore, ES need to be preserved to avoid such dilemmas or they will not still be considered for free. This is why scientists employ political and economic arguments to argue for the preservation of ES because economic aggregation figures may well serve as an indicator for raising attention for preserving them and show the need for monetisation of ES (Fongwa and Gnauck, 2009a). ES are usually considered as free goods, especially the less tangible services. Therefore giving them value is very challenging as they are under-estimated as almost valueless. One may think that payment schemes for ES can be an innovation for preserving them, but innovation can only be achieved, if value is added to them. To put an effort to link payment for ES with people's real preferences and willingness to pay (demand side), and on the other hand development of proper instruments to change resource management practices that delivered the demanded ES (supply side) may be very challenging (Fongwa and Gnauck, 2009a). This is due to the fact that it is not possible to get a representative Willingness To Pay (WTP) for potential impacts that reduce ES (that is preference about the society WTP), which correlates to an individual heterogeneity. But, representative WTP has led to the introduction of tax policy of equal opportunities that can not lead to the sustainability of ES because of free rider and 'Tragedy of the Commons' problems. "Tragedy of the Common" is an economic concept that means the consumption of public goods and services or providing externalities to the public at the expense of the society (Hardin, 1968). Therefore, the valuing of ES directly facilitates the development of strategies towards achievement of sustainability. Their values can be useful within the context of well-informed strategic decision-making to facilitate a better representation of the three pillars of sustainability such as follows (Slootweg and Beukering, 2008):

- Financial (Economic) sustainability of environmental and resource management
- Social sustainability by facilitating participation of stakeholders and by highlighting and addressing equity issues on ES to enhance their welfare
- Environmental sustainability by providing better insight in the long and short term trade offs of investment decisions on ES

Monetary value for ES might be assigned zero, but it is not correct and misses an opportunity to harness its value for a viable and profitable incentive-disincentive system that would guarantee the longevity of natural ecosystems. This is because without appropriate MES, the degradation of them may continue unabated and the lost of their services may have to be artificially replaced with expensive technology and at high collateral costs. Today, there are a lot of ongoing debates within the scientific community on the valuation of ES that need to be preserved (MEA, 2005; Banzhaf, 2005). Some literatures such as Wunder (2005) and Friends of the Earth International (2005) may argue that market-based approaches can not solve ecosystem degradation problems. Especially, as the poor in most developing countries depend mostly on natural ecosystem for their livelihood (World Resources Institute, 2008; MEA, 2005, World Agro-forestry center, 2007). This has led to some belief that it will be difficult to determine appropriate market-based arguments for business development for preservation of ES. But, it has been established that of all the jobs in the world, the majority have either a direct or an indirect relation to ES. Therefore, the possibility of targeting valuation for them does hold a potentially viable business model that can guarantee their preservation without undue degradation or free rider problems. Since it is a human right and ethically right that everybody has access to a healthy environment and their deriving services, by extension, it is also imperative that they should be preserved. This has already led to the introduction of incentive schemes such as market-based schemes for the development and conservation of them. Market-based schemes can be appropriately encouraged as an avenue for the development and promotion of measures to preserve ES. Payment and trading schemes for ES are emerging as world-wide small-scale solutions, where one could acquire credits for activities such as sponsoring the protection of carbon sequestration sources or the restoration of ES providers (Bradford et al., 2007; Daily, 2006). An example is the case of New York City and its water supply, where businesses discovered that investing in ES pays real dividends and are more cost effective than construction of water treatment plants (Shwartz,

2000). Therefore, establishing appropriate estimation approach for determining the value for ES can be a way forward to encourage business development for preserving them.

2.3 Business Development for Preserving Ecosystem Services

Business development and new business creation is the practice of initiating, organising, and developing new business opportunities (Krishnam, 2006). Business functions focused on strategies, creating strategic partnerships and long-term relationship with its stakeholders (Sustainability). Sustainability is an internationally accepted vision and has long been an important part of the medium- and long-term strategies of successful businesses (UNCED, 1992). The term sustainability was originally adopted in Agenda 21 of the UN program on sustainable development, before the widespread of the term sustainability, sustainable industries and sustainable economy become common. Therefore, sustainable business development implies a corporate policy that takes into account business success, environmental impacts and relations to the surrounding community. This requires putting them at a balance footing towards achieving long-term targets/objectives (BSDglobal, 2007). Corporate environmental and social policy is not an end in itself, but encourages success in business. Innovative, well-managed businesses use environmental management systems as a way to make efficiency gains. At the same time, the growing awareness of social and ecological responsibility on consumer and financial markets motivates businesses to work for greater sustainability. Consumers will like to buy from producers, whose processes are environmental friendly, thereby also determining the direction in which investors would like to put in their finances.

The business community believes that the main purpose of business development is to make as much profit as possible (Steiner, 2005), but modern business development do not only based on profit motivates (Kotler and Nancy, 2008; Allan, 2005). The sustainability of businesses is more important. If, this holds, then the concept of ES based on the "indirect value in use" may be an argument for their business development (Pagiola and Platais, 2003). That is businesses can be investing in processes that would improve their environmental performance, provide goodwill (intangible assets to business) and Corporate Social Responsibility (CSR). Preserving and restoring natural landscapes and the associate services they produce to humans offer business opportunities that can be harnessed. Therefore the values of these markets, especially for none- tangible services have to be made attractive to encourage stakeholders to be engaged in their development. There are some literatures with MES (Bradford et al., 2007; Hartig and Drechsler, 2009; Fongwa et al., 2011). These markets developed as a result of problems on the landscape that threatened the environmental quality leading to protection strategies for them, which business development for ES is an option. For example, the increase in emission level, especial atmospheric CO_2 has led to increase global warming potentials as a result of combustion of gases and land use changes (IPCC, 2003). This has led over the past years to dramatic changes in rainfall pattern, rise in sea and ocean levels, and disruption of other processes such as thermohaline current.

2.3.1 Market Structures for Preserving Ecosystem Services

The proper function of markets depends on institutional, cultural and legal capacity (Friends of the Earth International, 2005; Greenspan-Bell and Russell, 2002; von Weizsächer et al., 2005), especial markets for environmental management that demand a high degree of regulatory frameworks. In many countries in the world, these capacities are weak, particularly regulatory frameworks, which can lead to arguments against MES. Therefore a presentation of market structure for preserving ES can support the argument for MES. Figure 1 presents the linkage between market participants and a general market system (competition or regulation or both) that goes to influence market interaction in delivering of goods and services to consumers. It shows a linkage of different stakeholder participating in the market such as suppliers, consumers, government and investors, while the market behaviour is based on competition and regulation. The suppliers supply services to other market participants (consumers) within the market based on competition and some regulatory framework from the government (Government participate in the market to regulate it). On the other hand, the investors participate in the market to finance suppliers and consumers of services, and also targets to achieve government policies. The establishment of Public Private Partnerships (PPP) to support projects for preservation of ES with different capacities requirements, in which they may be passive and active partnerships. For example, for financing projects on ES, passive partners generate direct capital from government agencies to private entities in the form of grants, loans, equity or insurance. Active partners are those that capital and other resources are contributed to them by private entities for a venture, with government providing grants, supportive regulatory policies or legal protection and enforcement. Many companies are spending a lot of money to reduce environmental damages such as coal firing power plants and many processing industries. The Organisation for Economic Co-operation and Development (OCED, 2007) argues that as long as environmental harmful activities are less costly or more profitable than eco-friendly practices, then people and businesses would only cheat or make only token contributions on environmental protection.

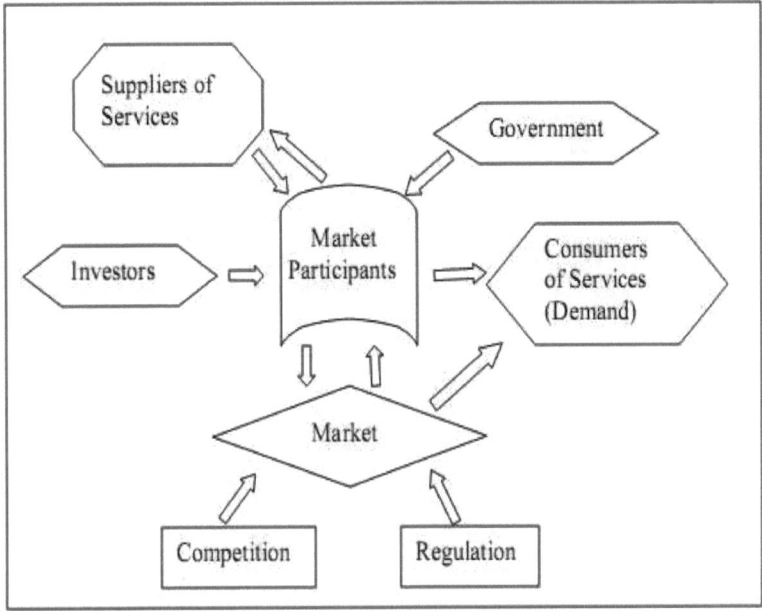

Figure 1: Market Framework (Fongwa et al., 2011)

This would cause them to continue with their business activities as usual, but at the cost to society than favouring environmental friendly schemes. This argument supports the fact that MES need to be back up with appropriate legal and institutional capacities, and also reducing transaction cost to make it attractive for stakeholders participation in preserving ES. Figure 2 present a market framework with a management company and the government as the only main market participants that determine the structure of MES (demand and supply for ES). It shows the linkages of the management company with the consumers, suppliers and investors, and the market. The government participate in that market to provide the rules and regulation for the functioning of the management system. The company manage the demand and supply of services, and investment, thereby controlling the consumers, suppliers and investors in the demand and supply of ES. The management company is then being controlled by the government. Many companies are using management companies in addition to mandatory compensation to offset damages to ecosystems, which increasingly more companies are becoming interested in the potential public relationship benefit of voluntary offsets to encourage the preservation of ES.

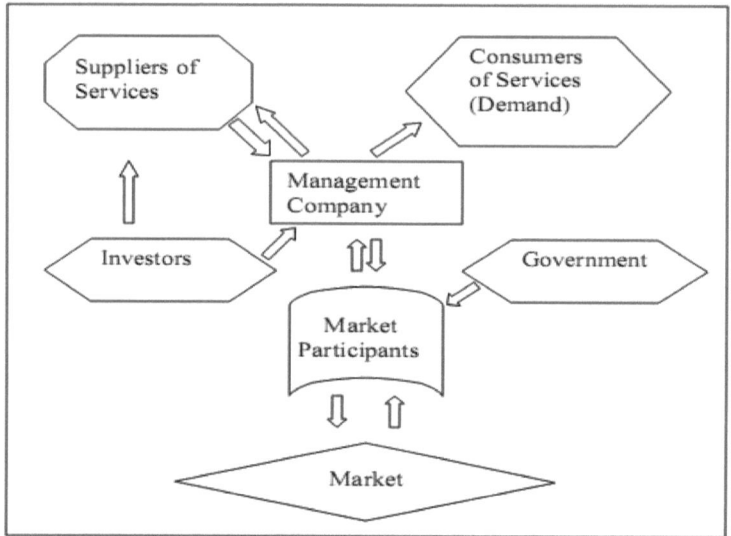

Figure 2:   Market Framework with a Management Company (Fongwa et al., 2011)

Some mainstream investors are beginning to look at biodiversity offsets as a business opportunity, as well as an indicator of good corporate governance in the companies they want to invest in, given power to these management companies to grow (IFC, 2006). Some investors are also seeing the business potential of these management companies for ES and are making direct investments in them. But the problem is that they may not give incentives and motivation for some companies and individuals to make personal efforts like development of cleaner technology or resource saving strategies to reduce damages to ES. These management companies can also gain monopoly power and turn to be acting on their own best interest that may be detrimental for societal interest leading to a hyper reduction of ES. This means that MES have to be clearly separated from conventional markets and encourage multiple stakeholders' involvement as market participants. Figure 3 presents a market framework with multiple participants that include the management company, suppliers, government, investors and the consumers having multiple options for demand and supply of ES. They may depend entirely on the management company or use other market participants based on market conditions, especially if they have perfect information about the most appropriate source of their demand or supply of ES.

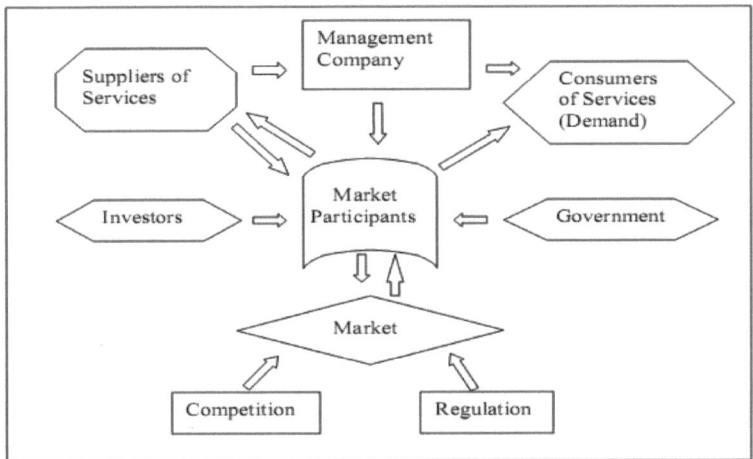

Figure 3: Market Framework with Multiple Participants (Fongwa et al., 2011)

The integration of multiple stakeholders as market participation can lead to reduction of transaction and the attainment of Corporate Social Responsibility (CSR) that need to be foster through increase in communication and information to the market participants (Allan, 2005). Increasing communication and information on MES can change the behaviour of market participants on the consumption of ES and contribute towards environmental protection and sustainability. There are basically three well recognised emerging MES, which are carbon sequestration, water and biodiversity markets. But understanding the science involved in the supply of ES and allocating their appropriate demand, one can classify MES under these 3 forms of markets. Therefore understanding how these markets are developed would offer the possibility of creating additional MES, which can be an innovation. For example knowing the extend to which reforestation activities improve the water quality for downstream users and the socio-cultural and political constraints on land use management, and recommendation of policy tools can encourage MES. MES are emerging in many countries in the world with remarkable examples in American countries like Brazil, Panama and USA (Bradford et al., 2007; Houck, 2002; Huber et al., 1998).

2.3.2 Markets for Carbon Sequestration

There have been developments on market niches and a great deal of speculation in both voluntary and regulatory markets defined by the Kyoto Protocol and the European Union's Emissions Trading Scheme (ETS) (Lauterbach, 2007). Under the Kyoto Protocol, protected existing forests is not an accepted source of carbon credits, but afforestation and reforestation projects are potentially eligible to function as a source of carbon credits under the Clean Development Mechanism (CDM). Other CDM for CO_2 sequestration is geo-sequestration by soil and algae lakes that are practiced mostly by big power plants (Stephens, 2006; Benneman, 2003). The buyers of credits in these markets include industries, governments, None Governmental Organisations (NGOs) and individuals. These markets are global market and offering regional areas to enter the market. This regulatory market is supported by existing government requirements prior to a country's ratification of the Kyoto protocol. It has entered into force in the EU Emission Trading System (ETS).

Emission trading is provided by "article 17" of the Kyoto protocol and it allows the commercialisation of "Assigned Amount Units" (AAU´s). Other units that are tradable under the scheme are; a Removal Unit (RMU), Emission Reduction Unit (ERU) by joint implementation, and a Certified Emission Reduction Credit (CERC) generated from "Clean Development Mechanism" project activities (UNFCCC, 2005). The EU amendment of Directive 2003/87/EC; Directive of 27th October 2004 establishes a scheme for "emission trading" in respect of the Kyoto Protocol's Project Mechanism. In such schemes emitters of CO_2 gases can reduce their overall levels not only within the internal processes, but also externally by purchase of carbon offset units or "credit" that is generated by recognised carbon sequestration or emission avoidance activities (Bradford et al., 2007). To support the growth of this trading scheme, regulatory and voluntary markets have been established to ease the exchange of credits from supplier to purchasers leading to opportunities to generate potential significant profits from the sale of carbon credits. This market is still young; despite rapidly growing one may think there are some uncertainties, especially as they relay on natural conditions that are dynamic.

2.3.3 Markets for Water

Markets for water quality and quantity are just like carbon sequestration market, but it is limited to local or regional scale (Johnson et al., 2001; Friends of the Earth International, 2005). They can be linked with the Cleaner Development Mechanism (CDM) market to have a broader scale. Forest cover in a watershed can be managed to provide a big stream of hydrological service that benefits a number of users such as downstream users of water resource in a watershed. They may provide the following services; regulate water flow by controlling flooding and seasonal flow, and improving water quality by increasing dissolved oxygen levels and reducing erosion, sedimentation, pathogens, eutrophication and chemical contaminant loading in water (Chapin et al., 2002; Landell-Mills and Porras, 2002; Bruijnzeel, 2004; De Barry, 2004). Household and industrial contaminants, pesticides and fertilisers from farmlands and algae blooms in water are other aspects that are problematic to water quality. They are likely related to significant increase in the concentration of phosphates and nitrates leading to eutrophication in water bodies. Therefore land managers have to determine whether it is economically desirable to manage their land to supply watershed services by understanding the existing demand. Local market may be particularly desirable for small holders like upstream neighbours being compensated by downstream neighbours in the form of tangible or intangible assets like donated labour or contributing in the reforestation or afforestation of the area around the head waters (Bradford et al., 2007). This means market instruments can provide the provision of water related ES in terms of quality and quantity. Therefore such instruments can stimulate the demand and supply of hydrological goods and services, which need the determination of market structures of water resources for preserving ES.

Market-based approaches also provide interested purchasers to buy water quality "credit" either to promote waste water treatment concerns or maintain forest cover that supply fresh water. It can also be through nutrient reduction in water (Houck, 2002). A nutrient reduction trade involves an exchange of effluent control responsibility between discharge sources. The control responsibility is expressed in terms of "allowance" or "credit," which specifies the quantity of effluent that a discharger is allowed to release. An exchange of allowances or credits does not increase the overall effluent discharge. Increased discharges by one source are offset by decreased discharges by another source. It also places a cost on the source's decision to continue to discharge effluents. In a trading system, the cost is the price to purchase allowances from another source. Within a properly operating trading system, the financial incentives for discharges to reduce costs drive measures to search for more effluent control strategies. These types of business developments can encourage markets for removal of impurity in water bodies like salinity (salinity credit) and organic matter, hence reducing eutrophication problems in water bodies. Credits for such activities can then be sold to companies or individual, whose activities lead to increase in organic matter and impurities in water. This can help them meet regulation measures and threshold limits. There are some companies along the river Spree in the BR that remove organic matter in water that are funded by the state. Therefore the proper way to organise water resource management in the watershed is by bringing all the stakeholders to cooperation for benefits and cost of damages sharing based on pareto optimality strategy.

2.3.4 Market for Biodiversity

Markets for biodiversity are just like that for water and carbon sequestration, which is also a MES that is growing rapidly with typical market structuring being bio-prospecting. This is the practice of mining natural genetic and bio-chemical resources in search for pharmaceutical, therapeutic, or agricultural products (Castree, 2003; Artuso, 2002). This type of market has led to the discovery of new drugs and chemicals offering a potential for pharmaceutical industries to enter this market (Gilbert et al., 2003). Biodiversity comprise of a lot of services such as material goods (timber, fish, fruits, medicines, rubbers, sources of bio-energy) and environmental services (generating and maintaining soil, converting solar energy into plant tissue, sustaining hydrological cycles, storing and cycling nutrients, controlling the gaseous mixture of the atmosphere, regulating and climate) (Myres, 1996). But these goods and services from biological diversity in most region in the world are being threatened or rapidly diminishing leading to measure to preserve them that has encouraged the growth of markets for their conservation (Vedder and Wright, 2003).

Species and wetland mitigation are another MES that comprise of projects that create or enhance wetlands or endangered species habitants for banking (Wilkinson and Kennedy, 2002; European Science Foundation, 2007; Fox and Nino-Murcia, 2005). These banks then sell credits to developers to satisfy their permit requirements for destruction of species during their developmental activities. This means mitigation banks can pull private industry into environmental preservation business, thereby giving developers the option to buy credits or share in the conservation fund's mitigation banks to offset their ecological damage that their developmental activities cause on the environment. Conservation bank credits are available to public and private enterprises that are currently having projects that impacts high elevation wildlife habitats (ten Kate et al., 2004; Jenkins et al., 2004; Johnson et al., 2001; Wissel and Wätzold, 2008). The habitat conservation banking focuses on providing a mitigation solution to unavoidable impacts to habitat, other sensitive and endangered species. The bank operators are permitted to sell habitat credits to those who need to satisfy legal requirements associated with environmental impacts. These markets are regional, but can be integrated in the CDM to be global. That is companies in countries that have ratified the Kyoto protocol can be sponsoring biodiversity conservation projects in threatened biodiversity regions (hot spots) to integrate them with the emission reduction scheme. These market types can encourage fund derived from successful commercialisation of such products to be put back to pay for conservation of threatened ecosystems or compensating indigenous peoples, nations, and related stakeholders for access rights, and in some cases to reward indigenous people for their intellectual capital (Bradfords et al., 2007). Markets for biodiversity banking are being developed in countries like Brazil as a "Tradable Rights" that is used to recognise the value of biodiversity in an area (ten Kate et al., 2004; Wissel and Wätzold, 2008). The rights can be traded to develop or not to develop an activity that impact ecosystems. It does not take into consideration only particular species or products, but a location or a piece of land. The rights allow for compensation with some tax incentives or some tax break that are transferable.

In addition, Market-based mechanism for biodiversity management can established eco-labelling and certification schemes to distinguish products and services by their social and environmental performance. The back born to such schemes is that consumers who want to support the maintenance and development of ES would prefer to buy or even pay more for certified goods and services. It may even influence the behaviours of consumption of goods and services (Kotler and Armstrong, 2001; Kotler and Nancy, 2008). A number of certification schemes have gained consumer recognition world wide and market share for some certified products in some markets are gaining more grounds than that of uncertified products. Certification schemes have also been seen as a good tool for encouraging and recognising environmental and social best practices in a range of business sectors like agriculture, forestry, tourism and even financial services. Different forms of agricultural certifications are growing worldwide like the International Federation of Organic Agricultural Movements (IFOAM) found more than 31 million ha of farmland under organic management world wide (see http://www.IFOAM.org). Many food and agricultural companies are also promoting and buying certified products, and labels provide a guide to them not to be confused with uncertified products (Willer and Yussefi, 2006).

Sustainability of forest resource management in the forest sector just as the agricultural sector is increasingly tested and validated through certification by independent organisations. There are regional and national organisations for sustainable forest management developed by the Forest Stewards Council (FSC) as well as the Programme for the Endorsement of Forest Certification schemes (PEFC) in Europe, USA, Canada, Australia, Brazil and Chile (UNECE, 2006). Recreational use of forest and other ecosystems is growing rapidly due to the expansion of domestic and international tourism to manage the growing pressure on ecosystems arising from increasing number of tourists. This has increased the value derived from nature-based tourism, tourism guidelines, certification and accreditation schemes, which have been (or are being) developed. The United Nations Tourism Organisation has published a code of ethics for tourism in 1999 (see http://www.world-tourism.org). There is also the final report of the World Tourism Summit held in Quebec in 2002 that recommended on the development of guidelines on certification schemes for ecotourism (World Ecotourism Summit, 2002). The Convention on Biological Diversity (2004), in a partnership with the tourism industry has developed "Guidelines on Biodiversity and Tourism Development". There are also tours operators' initiatives for sustainable development, which is developing environmental guidelines for hotels, resorts and tourist attractions located in or near biodiversity hotspots. Sustainable Hotel Sitting, Design and Construction have also been adopted by many large hotel chains, and there are also tour guides (Conservational International, 2007). Recently, the Sustainable Tourism Stewardship Council (see http://www.rainforest-alliance.org/tourism.cfm?id=main) has been created and may offer a route to harmonisation of sustainable tourism approaches. These markets are local or regional; for instance in the Spreewald region where the BR is found, markets for certification scheme are growing with the "Spreewald Logo" as the label to identify certified products from uncertified products with many firms participating in the scheme (see http://www.spreewald-erlebnis.de).

Many researches are going on "Transferable Development Rights", instruments for sustaining ES by designing market structures on certification process with defined exchange and monitoring rules that can be enforced by regulation to function. A typical example is tradable permit market for development of ecosystems, which is in progress with many researches going on to develop this type of market (Eco-trade, 2009; Carson, 1991; Wissel and Wätzold, 2008; Tietenberg, 2002). The idea behind such market is to develop an ecosystem management scheme that can create new rights or liabilities for the use of natural resource (preserving and maintaining ES). Then allow businesses to trade on these rights and liabilities. This means a company can either transfer its liability of an environmental damage or its rights of the use of a natural resource to a management scheme, which it is then tradable. Some experiences with such markets have shown significant reduction of the cost of protecting the environment and/or maximising the value of resource use. These MES are still young that need incentive and motivation schemes to make them grow and to engage a broader number of stakeholders in their involvement. These MES can be however established into 3 classes; permits, certification and conservation credit/banking. Permits can be used for water and carbon sequestration and conservation banking/credit for biodiversity. Certification can be used for all the 3 markets of carbon sequestration, water and biodiversity. Permit, certification and conservation credit/banking are used in the next section on methodology (see section 3) as one of the framework for constructing the unified model for management of ES at the landscape scale.

2.4 Trading Platforms to Establish Markets for Preserving Ecosystem Services

Trading platforms for ES are being established at both the international and local level to encourage the development of MES, especially to involve the private sector. Allocating ES through a price mechanism implies that those who are willing to pay a price for ES meet those willing to sell at that price within a specific period of time and place (Begg and Ward, 2004; Mankiw, 2008; Smith and Begg, 2000). But it is sometime not usually the case with trading platforms for ES, which are mostly based on incentive and motivation schemes to achieve regulatory targets. Therefore a potential trading platform for ES is a clearing house mechanism based on the model of the ecosystem marketplace (Daily and Ellison, 2002; Van Bueren, 2001; MEAs, 2005; Carson, 1991), where demand and supply for ES can meet. That is consumers and suppliers of ES can come together to negotiate. Thereby a platform where reliable information on projects for the provision of ES that private companies, foundations, Inter-governmental organisations and NGOs can either finance or undertake. These platforms can significantly reduce transaction costs for suppliers and consumers of ES, which involved biodiversity, carbon and land-related services that can be traded.

Insurance companies that understand the function of ecosystems and the services they produce can also develop premium policies on highly risky areas. There are such types of business innovations already growing among ecosystem developers and even some insurance companies are now creating risk policies on assets that are really critical to environmental damage (Munichre, 2000). They are developing proactive environmental policies, such as lobbying for increase on global climate change research and the development of ES. This underscores the business risks and opportunities perceived by the insurance industries, especially if the pay-outs for environmental disasters related to climate change become significant. These insurance companies have begun to look at how degraded ES are going to affect their market bottom line and are encouraging business development for ES. To foster MES and the trading platforms to grow however, many international organisations are undertaking various initiatives. Several United Nations (UN) agencies and programs have been undertaken to conduct studies and field work on MES like the Bio-trade developed under the UN Conference on Trade And Development (UNCTAD), joint program between UNCTAD and the Earth Council Institute (Bio-trade, 2008; UNCTAD, 2009). World Bank's activities for fostering market-based instruments for ES and the Inter-American Development Bank undertake to work in the field of eco-tourism as well as to identify new avenues for investment in forest and biodiversity conservation, and rational utilization of resources (IFC, 2008a and 2008b, IAB, 2005). Multilateral Environmental Agreements (MEAs, 2005) provide tested mechanisms for setting up trading rules for ES with two possible trading platforms under the scope of the Rio convention (Paquin and Mayrand, 2005). The creation of funds to collect finance from local, national and international organisations and to fund institutions that carries out projects for developing ES. Therefore analysis of market approaches for ES can provide multi-stakeholders with guidelines on business development for preserving ES.

Multi-stakeholders' engagement as market participants for ES can encourage trade creation leading to competition. Competition means more choice to participants, which can lead to social welfare of society because eco-friendly practices can become cheaper, thereby encouraging increasing number of participants to be involved in their preservation (binary theory). This is because market creation and widening for ES is a potential than market concentration or diversion from them. Their involvement can also lead to cooperation for development of MES based on pareto optimality that can be achieved through CFP. This is applying a strategy to obtain the best outcomes. Therefore, one can say a state is optimal, when it can produce the best outcome between the maximum and minimum levels. When there are mechanisms with appropriate redistribution of wealth, a Pareto optimal resource allocation can also be attained through a competitive market schemes. Wealth distribution within individuals is essentially a moral question. But through pareto optimality allocation, gainers in principle will compensate the losers based on the condition that they were participating. Rawl (in Rubinstein, 1998) has argued on this based on the "veil of ignorance" of no knowledge of inherited circumstances in which it would be most reasonable for a community to care about the worst-off. This leads to the idea of ownership because people have different incentives for ownership like for example the case of the sole proprietor that spent a lot of time in his business because he is the sole owner (Zoltan and Gerlowski, 1996). This means given ownership rights to individual would prevent them from taking actions that can directly affect others, especially if the strategy of pareto optimality applies. Therefore CFP can provide an incentive that can be use by stakeholders to initiate business development for ES as well as provide financial resources for their development.

2.5 Community-based Financial Participation

The concept of CFP was newly developed within the framework of this thesis research. It is an extension of the concept of Financial Participation (FP) that had its origin and relevance in the late 80s (EFES, 2001). The thought of FP is how practice of Employee Share Ownership (ESO) could join with participatory management that can have an impact on economic and social dynamics of corporations. Since then, many research initiatives on FP have been carried out around the world like the European Commission sponsored project on the Promotion of European Participation in Profit and Enterprise Results (PEPPER project), Employee Share Ownership Plan (ESOP) and FP exists in many countries in the world today (Lowitzsch, 2007). But CFP have not exist before, only within the framework of this thesis research that it was developed with some few literatures (Fongwa and Gnauck, 2009b and 2009c; Fongwa et. al., 2009 and 2010a). The idea of CFP is that FP should not only remain within corporation but can be extended to the community-based level to have a broader meaning (Krishnam, 2006). In 1924 and 1974, the USA economy was characterised by an abysmal lack of purchasing power, leading to an extreme low rate of productivity, confrontation within agents of society and lack of capital growth and expansion (Lowitzsch, 2007). Interest rates were also at all time high, and few banks were willing to lend in any event. The stock market was at its lowest since 1929, and public participation was non-existing. This had encouraged the idea of the idea of FP, which can then be extended to CFP.

CFP is a pareto optimal strategy requiring community stakeholders to pull financial portfolio for development of community resources, which will allow them to have ownership rights and benefits from the development of the resource based on benefit sharing. Figure 4 presents a conceptualisation of CFP that can be used for business development for ES. It shows the linkage of financial portfolio for contribution to equity from shareholders and employees that can be involved in the supply and demand of ES for their development. On the other hand, contribution to equity can also come from financial institutions like banks, consumers and customers of ES (Households, Firms, Suppliers and others), other businesses and institutions in the community (Insurance companies, mechanical companies and others) and government (Fongwa and Gnauck, 2009c). Therefore, capital acquisition for preservation, provision and management of ES can come from the various contributions to equity and also from loan, if credit is being taken from financial institutions for the provision of ES. This is only in the case of high Net Present Value (NPV) investment projects. All the various stakeholders that can potentially invest in the provision of ES would participate in the sharing of the benefits that arise from the management of ES (profits) in terms of their contribution. The benefits can be monetary or in terms of other resources for all the players who contribute in the portfolio and none contributors can be left only with the functional benefits in which they cannot be excluded from them. The contribution to CFP may not only be liquid asset, but fixed and current assets to enable everyone in the community to have the possibility to contribute to the portfolio.

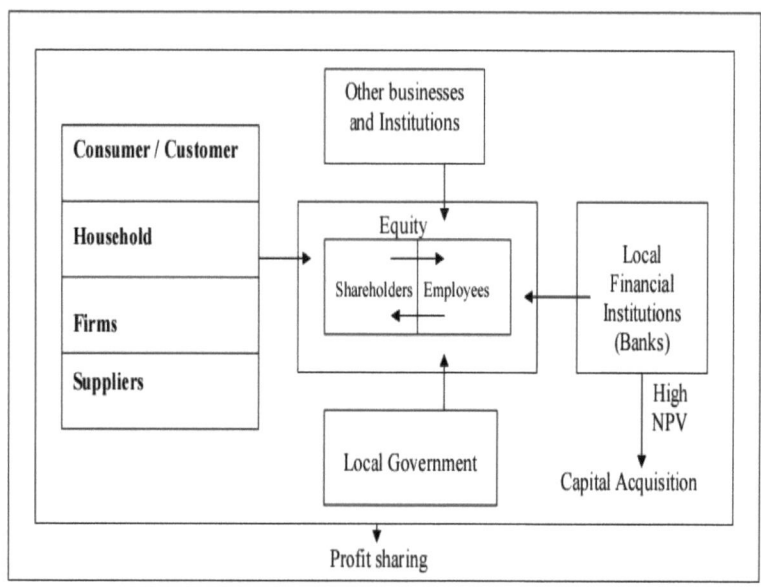

Figure 4: Community-based Financial Participation (Fongwa and Gnauck, 2009c)

Therefore for MES, landowners for instance may offer land or community stakeholders assisting in tree planting in upland areas, other stakeholders may be providing fixed assets and even current capital stock such as other form of labour like working in plants that reduce impurities in water and many more. This would potentially increase the capital stock for business development for preserving ES. Then, all stakeholders can participate in the benefit of the business according to their contributions (profit sharing). Therefore this concept can be used to potentially provide huge financial portfolio for achieving environmental objectives, especially for investment in new business development initiatives such as emerging MES. The establishment of relationships between different sets of ecosystem components (component of functions and processes that providing ES) with Multi-actors (Ma) and their Trade-offs (TO), one can understand and determine potential stakeholders for financing the development and maintenance of ES through business development. This is the concept of this thesis that leads to the construction of the modelling framework as a unified model for management of ES at the landscape through business development. That is a landscape type (L) need to be identified, which comprise of many ecosystems and their components that provide ES (see table 1 and 2 of classification of ES). Therefore for determining potential stakeholders for a CFP as incentives scheme for business development for preserving ES would depend on the economic TO of multi-actors demand or supply of ES. Development and maintenance of them requires an estimation system of the Value of ES (VU), which can provide ground for decision options. A multi-dimension analysis of the VU to stakeholders and their TOs with Ecosystem Component (COMP), can make it possible the determination of circumstances under which potential stakeholder can be willing to join a CFP for business development for preserving ES. This requires establishment of relationships and cooperations with strategic partners based on pareto optimality. Therefore the VU needs to be incorporated in analysis of business development options for ES. Figure 5 presents the concept for preserving ES to incorporate a multi-dimension strategy of the VU and the options for preserving them indicated by arrows. It shows on the one hand the linkages between ecosystems, their value that can lead to business development for preserving ES. On the other hand, stakeholders in ecosystems can establish a CFP for the development of business for preserving ES.

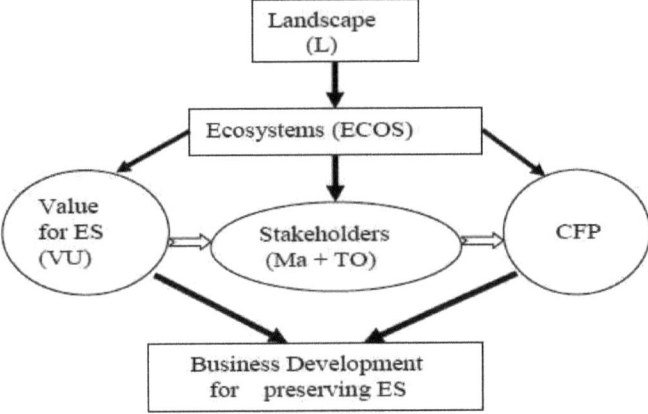

Figure 5: Concept of Business Development for Preserving Ecosystem Services (modified from Fongwa et al., 2010b)

3. Methodology

The methodology is a modelling and simulation framework with Petri nets. It is established as a computer based model for estimating ES for business development as a means of managing them at landscapes scales that can lead to nature protection. That is modelling multi-agent relationships (multi-actors and ES at particular landscape and ecosystem) for business development strategies for preserving ES. They are modelled as discrete events using graphical-based and mathematical approaches with Petri net (Eike et al., 2001; Löscher, 2009; Girault and Rüdiger, 2003). The whole system is modelled step by step (discrete) (Zeigler, 1976; Sajoughian and Cellier, 2000; Koji et al., 2007). This is based on multi-actors activities on the landscape that demand or supply ES, and Service Antagonising Unit (SAU) and Service Production Units (SPU).

The methodology is structured with section 3.1 presenting the constructions of Petri nets and description of their relationship of the multi-agent interaction for ES, management scheme and CFP for development of MES. They are coupled as unified nets for business development for preserving ES. The Petri net framework also comprises of creating meta data set for data mining and manipulation that can support multi-dimensional analysis for ES for decision support system on balancing them on the landscapes scales. Therefore data sampling and aggregation procedures were integration in the modelling framework for input data for running simulation. They are presented in section 3.2. The methodology ends up with section 3.3 presenting parameter for estimating the value of ES. This can ease the interpretation of results for decision support systems, especially for market-based strategies for balancing ES at the landscape scale.

3.1 Petri Net Modelling Framework

A Petri net is a graphical-based and mathematical tool for modelling and simulation of flow systems (Eike et al., 2001; Löscher, 2009; Girault and Rüdiger, 2003). Petri net can be used for continuous, discrete, instantaneous and hybrid modelling, and also support stochastic outcomes (David and Alla, 2005; Reisig and Rozenberg, 1998; Wil et al., 2003; Girault and Rüdiger, 2003). In table 4 types of Petri nets and their characteristics are presented. It shows that the coloured Petri net can be constructed for stochastic continuous/discrete/time coloured Petri nets. The Petri net class considered here is a stochastic coloured Petri net constructed for Discrete Event Simulation (DES). A Petri net consists of places, transitions, and arcs that connect them (Löscher, 2009; Reisig, 1983; Desel and Juhá, 2001).

Table 4: Types of Petri nets and their Characteristics

Types of Petri nets	Characteristics
Place/transition	Basic Petri nets, it is time-free, essential for qualitative analysis
Continuous	Model continuous state changes of systems, continuous firing of transition, essential for modelling continuous systems
Discrete	Continuous instantaneous state changes, step by step firing of transition, deterministic state changes, essential for discrete events modelling with defined states
Timed	Fixed time of state changes, timed firing of transitions, essential for modelling control systems
Stochastic	Random generation of probabilities of state changes, stochastic firing of transitions (probabilistic), essential stochastic processes
Hybrid	Combined continuous, discrete and time Petri nets, and also an instantaneous firing of transitions. It can also be with stochastic. It is essential for modelling hybrid and complex systems
Coloured	An extension of all the Petri nets classes given here, it can be continuous/discrete/time, hybrid and stochastic with the extension as coloured Petri nets, essential for analysis of large and heterogeneous variables, multi-variable modelling,

They are connected with regard to the sequence of flow. It can be parallel sequences or one sequence of flow depending on the modelling approach. Figure 6 presents the four main primary elements that made up a Petri net. They are places indicated by a circle symbol, transitions indicated by rectangular box, arcs/arrows indicated by directed arrow and tokens indicated by a round mark.

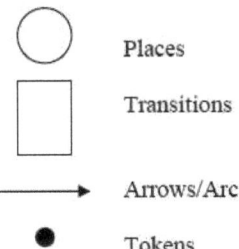

Places

Transitions

Arrows/Arcs

Tokens

Figure 6: Four Main Primary Elements of Petri Nets

Places are connected to transitions by arcs and transitions are also connected to places by arcs (Desel and Juhá, 2001). But a place and place nor a transition and transition cannot be connected in Petri nets. Places contain a current state of a modelled system (the marking), which is given by the number (different types if the tokens are distinguishable) of tokens in each place. Transitions are active components that modelled activities occurrence (the transition fires), thus changing the state of the system (the marking of the Petri nets). Transitions are only allowed to fire (set to work), if they are enabled. This means all their preconditions to fire are fulfilled. The interactive firing of transitions in subsequent markings is called token game. Therefore, Petri nets model actions based on changes in their local environment (state), which led to the notion of place/transition nets that was introduced in 1980 to distinguish this type of network model from networks without annotation and other network models (Peterson, 1981; Reisig, 1983). In general, actions depend on a limited set of conditions or restrictions that is called local environment or environmental rules. These rules determine the transition events (when fired). Therefore it is assumed that a certain action has to be provoked by some actors (multi-actors) to enable a transition event (represented by the rectangular boxes), which will determine the state of ES at that place. This means a behaviour of a transition exclusively depends on its locality, which is defined as the totality of its input and output objects (pre- and post- conditions, input, output processes), together with the element itself.

When a transition fires, it removes tokens from their input places and add them to their output places. The number of tokens removed/added depends on the cardinality of each arc. An example can be used to illustrate this. Figure 7 presents a Petri net of an ecosystem process. It shows that places are connected to a transition by arrows, and transition to a place. Places contain tokens. The numerical sign on one of the arrow is the cardinality of the arc. It determines the rate of flow from their input places to output place for an algae growth when the multiple resource kinetic fires. That is it removes one token from each of the 4 places (places containing radiation, substrate, nutrients and water temperature) and as a combination of 4 resources indicated by the weight of the arrow allows for the growth of algae.

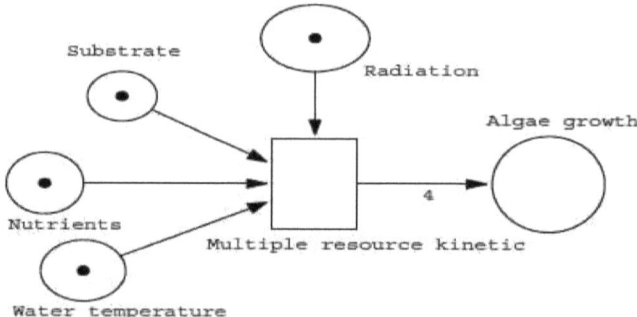

Figure 7: Petri Net of an Ecosystem Process of Multiple Resource Kinetics for Algae Growth (adapted from Gnauck, 1988)

3.1.1 Conceptual Model Building with Petri Nets

The model is constructed as a flow system that can be use to estimate the stock of ES after particular transition events. Figure 8 presents a Petri net model for flow of ES. It shows transitions as multi-actors and demand/supply for ES, which they control the behaviour of the model to determine the state at a place. The states in the model are MES, Service Units for ES and the Stock for ES, and ES as flows. The net is constructed in such a way that when the transition multi-actors fires a token is removed from the place "MES" and put to the place "Service Unit for ES". It can also move the other way back depending on the sequence of flow that must be specified.

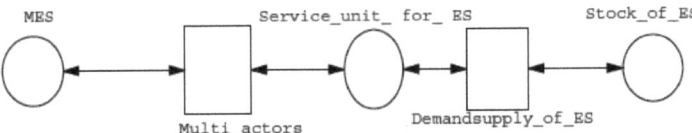

Figure 8:   Petri Net Model for Flow of Ecosystem Services

For instance, multi-actors demands or supplies of ES from "service units for ES" in the flow increase or reduce the "stock of ES", while market for ES is used to balance the stock of ES through a permit or certification or other marketing strategies. MES is imaginary and a logic place to "stock of ES". This is to be able to separate and model different actions caused by multi-actors or natural (like antagonising activities) that impact ES. The net is at a top level of generality that can be difficult in describing the system entities for generating the model behaviour for MES based on multi-actors relationships. Therefore the net is further extended to a lower level of generality, but with more complexity because of increase in places and transitions. Figure 9 presents an extension of the net with various multi-actors, the service units for ES that influence them through the demand and supply of ES. It shows multi-actors compromising of consumers, antagonisers, developers, migrants, suppliers and producers of services. It also contains the demand and supply units for ES, while the demand units for ES and the supply units for ES are also known as the Service Antagonising Unit (SAU) and the Service Production Units (SPU) respectively. A combination of them is called Service Units (SU). This can enable the estimation of the flow of ES over a landscape. Therefore, a graphical representation of Service Units (SU) as places for studying ES and multi-actors as transitions that trigger a change of the state of a place (ES) are established. ES are tokens that flow in the model, while textual and mathematical algorithms are used to describe the rate of flow for simulation. A huge variety of algorithms for the design and analysis of Petri nets have been developed to aid modelling processes. In designing a modelling framework for the concept, it is assumed that there are some ES in some places called service units for ES, which the actions of some multi-actor (transition event in the model) caused them to change their state (place).

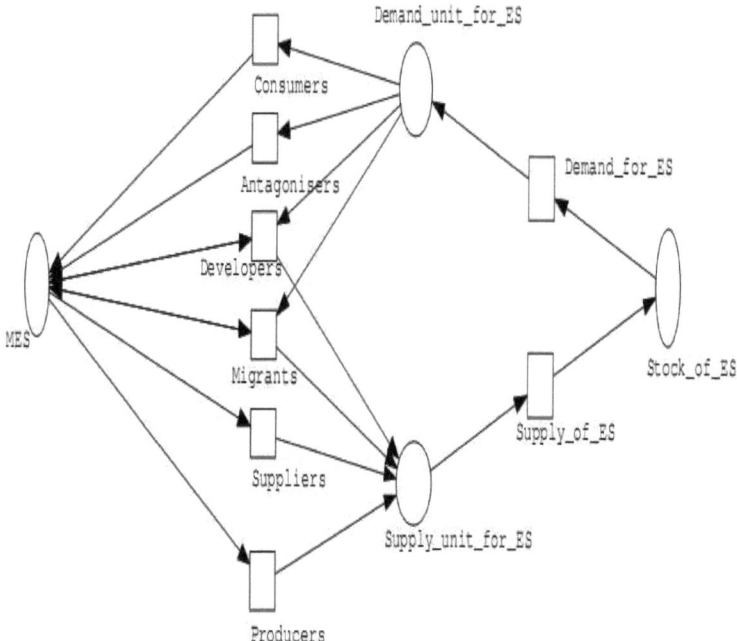

Figure 9:   An Extension of Petri Net Model with Multi-agents

The model contains two active parts, which are flow of ES and stock of ES. Changes in the state of a place are considered as flows enabled by an action. That is inflow for increasing ES (supply of ES) and outflow for decreasing ES (demand of ES) that determines the stock of them. Therefore in designing the model typical service centres for ES (Service Units for ES) are established for study of the interacting behaviours of multi-agents in the demand and supply of ES. This can enable the estimation of ES for market-based strategies that can lead to their business development. In the modelling framework with Petri net the attention is on causality among actions (Petri, 1962), and construct a sequence of computations that approaches the exact solution step by step (discrete approach). Then examining whether the limit of such sequence preserve certain behaviour of their components. The following specification of the model is; a landscape type (L) is identified as terrestrial ecosystems (ECOS) comprising of 8 ecosystem types, which can be further classified under the following ECOS:

- Agricultural ($ECOS_1$)
- Forestry ($ECOS_2$)
- Water ($ECOS_3$)
- Plants including green plants ($ECOS_4$)
- Urban ($ECOS_5$)
- Rural ($ECOS_6$)
- Urban-rural that is transitional ($ECOS_7$), and
- Settlements ($ECOS_8$)

Then each ECOS is further classified into Components (Comp), comprising of 3 of them such as land surface ($Comp_1$), water ($Comp_2$) and biodiversity ($Comp_3$) that are the markings, in which ES (tokens) can be modelled. The interactions of different multi-actors with these components are studied by simulation of their changes based on each actor's action. ES are tokens classified under the 5 types as follows: Provisioning (ES_1), regulating (ES_2), supporting (ES_3), preserving (ES_4) and cultural (ES_5). Then the typical services unit for the flow of ES are identified as a place that contains the markings with tokens for inflow and outflow of them (inflow to stock of ES). MES, Supply and Demand units for ES, and Stock of ES are assigned ES based on their 5 types differentiated by colours (Coloured Petri nets).

The places are controlled by the transition of action based on interactions of multi-actors based on their demand or supply of ES. This allows a simple firing sequence of Petri net to be assigned to these transitions. The construction of the modelling framework itself comprises only Comp and ES that are used in the specifications, while L and ECOS serve as attributes for data sources (Indicators for ES). They are essential for identification of background information for verifying input data (see section 3.2). The net is constructed as a generic model that simulates the flow of ES types or their categories for MES for a particular locality, region, and nations and even globally depending on the configuration of variables and data set in the places of markings. A management scheme for MES is introduced that provide strategies for preserving ES by their interaction in the internal and external market for balancing targets. The places in nets are management scheme, market scheme and external market scheme, and the transitions are market management and external market players. The flows are permits, certification and conservation credits/banks. The specification is that the management scheme can structure the MES using the market strategies like permits, certification and conservation banking/credits as market schemes to balance environmental performance on ecosystems. While external market players for permits, certification and conservation banking/credits can also demand or supply these market strategies. Figure 10 presents a Petri net of management scheme for MES that shows the places of flow of the various MS and transitions that controls them. It shows the places as management scheme, market scheme and external market scheme, which the transitions that fires to move the MS from place to place are market management and external market players.

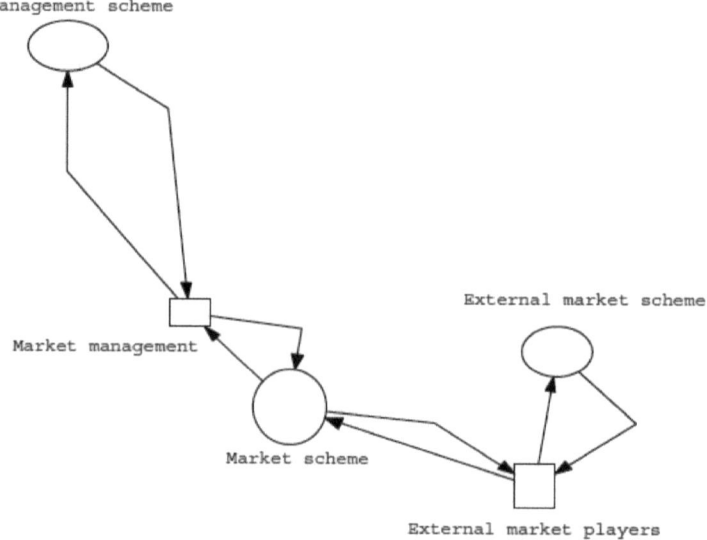

Figure 10: Petri Net of Management Scheme for Market for Ecosystem Services

Therefore for incentive and motivation schemes to support the proper functioning of MES, a CFP is integrated in the modelling framework. Figure 11 presents a Petri net framework of CFP for supporting the growth of MES. It shows that at a place, there is finance (Place Financing) controlled by different multi-actors (at transitions: local institutions/businesses, local investors, local government, banks and stakeholders), who can contribute to financial portfolio at the place contribution unit. The contribution then flow to the portfolio for financing business development for preserving ES, which the benefits from the business development flow back to the place "benefit unit". They are then distributed to the contributors. They might then decide to re-contribute to the portfolio or saved their finances.

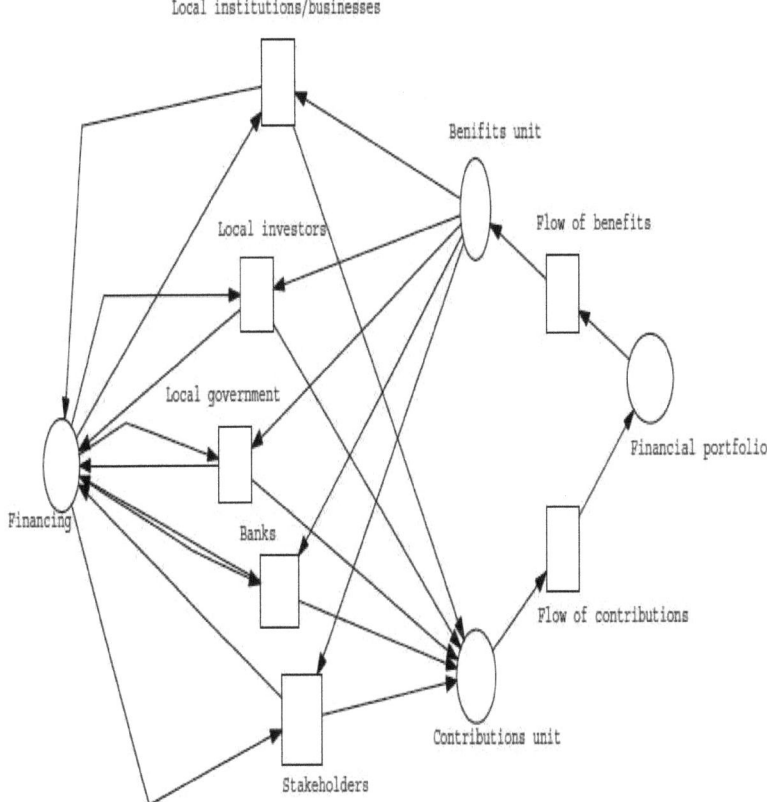

Figure 11: Petri Net of Community-based Financial Participation for Market for Ecosystem Services

3.1.2 Petri Net Relationships

In general a marking is a function of tokens to the set of natural numbers of $\mathbb{N}_m = \{0, 1, 2, ..., m\}$. If the Markings of a place S is n, then one can says S contains n tokens (Eike et al., 2001). When $n = n_i$ and $i = (1, 2, 3, 4)$, then the Petri net (see figure 9) contains the markings

Comp in the places MES, Demand/Supply units and Stock of ES expressed as n_{1-4} = Comp set of Markings with tokens (ES). That is n_1 is markings at the place MES, n_2 is markings at the place Supply unit for ES, n_3 is marking at the place Demand unit for ES and n_4 is markings at the Stock of ES. The markings are specified as follows:

Comp = {$Comp_1$, $Comp_2$, $Comp_3$} (1a)
ES = {ES_1, ES_2, ..., ES_5} (1b)
$Comp_1$ = {ES_1, ES_2, ..., ES_5} (1c)
$Comp_2$ = {ES_1, ES_2, ..., ES_5} (1d)
$Comp_3$ = {ES_1, ES_2, ..., ES_5} (1e)

Colours are thought of as differentiating a variable and/or data type (Girault and Rüdiger, 2003). Therefore they are used to differentiate the different types of flow variables (tokens) or markings. Since each place may contain objects of specific colours. Then for each Place (P) a colour domain or colour set cd (P) is defined. Other examples of colour domains are set of integers or Boolean values. The different markings and their set of tokens are differentiated by colours depending on the goal and type of investigation. For instance, if the goal is to measure all ES without differentiation, then all the ES and their categories will have one colour. But for measuring different categories of ES, then the categories are being separated by colours (coloured Petri nets). For example, if a marking $Comp_1$ with tokens of a set of ES (ES_1, ES_2, ..., ES_5) are to be estimated at different places of flow, then one colour of a set can be defined as "$COMP_1 ES_1$". Therefore the colour sets can be defined:

$$\text{Colours} = \begin{pmatrix} Comp1ES1 & Comp1ES2 & ... & Comp1ES5 \\ Comp2ES1 & Comp2ES2 & ... & Comp2ES5 \\ Comp3ES1 & Comp3ES3 & ... & Comp3ES5 \end{pmatrix} \quad (2)$$

This means n_{1-4} will contain the following colours in each of the places; {($Comp_1ES_1$ $Comp_1ES_2$ ...$Comp_1ES_5$), ($Comp_2 ES_1$ $Comp_2 ES_2$ ... $Comp_2 ES_5$), ($Comp_3 ES_1$ $Comp_3 ES_2$ ... $Comp_3 ES_5$)}. Thereby one place has 15 colours of the different tokens. Then the flows of tokens (colours) are from "Demand/Supply units for ES" to "Stock of ES" and vice versa depending on the direction of flow indicated by the arc in a particular sequence.

However, the behaviour of a marked Petri net with initial markings is a set of execution sequences, which allows the flow of marking into other markings through occurrences of transitions (Eike et al., 2001). A marking can flow in a place of other markings if it initially contains that type of markings (definition of coloured Petri nets). This argument allows the rule for a change of marking to be described. $Comp_1$ with a coloured token "$Comp_1 ES_1$", for instance at the place "MES" is enabled to the place "Supply unit for ES" or n_1 to n_2, if the flow (input) contains the definition of at least one of such a coloured token. A change of marking in a place occurs only when an enabled transition is fired that releases or accepts token(s). That is a token or tokens are removed from an input place and added to an output place without other places being affected. The input token(s) in the output place is directly reachable from the input place, therefore establishing an important relationship (Eike et al., 2001). The closure of the reflexive transition gives the overall reachability relation for markings. On the other hand these rules apply for the management scheme and CFP nets. This is an advantage of Petri net that one rule is defined and it can apply to others. For instance the colour sets management scheme is defined as follows:

$$MS = (MS_1, MS2, MS3) \quad (3)$$

Therefore colour set for MS is defined:

$$Colours = (MSMS_1\ MSMS_2\ MSMS_3) \quad (4)$$

While the net CFP has two tokens (a and b) of flow and a marking (fin), and they are represented by colours defined as follows:

$$Fin = (a, b) \quad (5)$$
$$Colour = (a\ b) \quad (6)$$

3.1.3 Formal Introduction of Net Relationships

Formally if the Petri net (see figure 9) is defined as N that described a triple (S, T, W) and S is the states (places) of the net N, while T the transitions and W the weights of the arc. Then sets of S and T can be defined on N for flow sequences through W as follows:

$$S = (s_1, s_2, s_3, s_4) \quad (7)$$
$$T = (t_1, t_2, t_3, t_4, t_5, t_6, t_7, t_8) \quad (8)$$

W assigns a number (in this case, only 0 or 1) to every pair (s, t) and (t, s), where $s \in S$ and $t \in T$ depending on whether an arrow or not leads from s to t or from t to s, respectively (Eike et al., 2001). For instance, if an arrow leads from s_1 to s_2 are assumed a and b, then $T' = \{a, b\}$, $W(s_1, a) = 1$ and $W(s_1, b) = 0$, where 1 stands for "arrow" and 0 stands for "no arrow". The real weight can then be expressed on the arc as a multiple, function or algorithm that determines its flow. The marking n can be described as a function satisfying the rules; when $n_1(s_1) = 1$, then $n_1(s_2) = 0$, $n_1(s_3) = 0$ and $n_1(s_4) = 0$, and when the marking n_2 as a function satisfying $n_2(s_1) = 0$, then $n_2(s_2) = 1$, $n_2(s_3) = 0$ and $n_2(s_4) = 0$ and so on till $n_4(s_1) = 0$, $n_4(s_2) = 0$, $n_4(s_3) = 0$ and $n_4(s_4) = 1$. If for instance n_1 is being enabled by a transition "a", then it occurs as far as the condition for accepting tokens are satisfactory or it rejects. If it accepts, then the resulting marking is n_2. Thus, $(N, n_1) [a > (N, n_2)$ is described. Intuitively, N describes a sequence of two transitions "a" and "b" relating to a communicating sequential process (Milner, 1989). Therefore the behaviour of a set (N, n_1) is determined through the transition rule of Petri net. N by itself has no behaviour (as no transition is enabled) and in the marked net (N, n_1), b can occur but not a. In table 5 a summary of firing sequences and formal net relations is presented that is the basic for its kinetics. It shows the corresponding input at a place when a transition fires. That is 4 places with 8 transitions, which fires to produce its inputs of Boolean values as is established in the formal relations. It forms an 8 x 4 matrix.

Table 5: Summary of Firing Sequence and Formal Petri Net Relationships

Places/Transitions	n_1	n_2	n_3	n_4
t_1	1	0	1	0
t_2	1	0	1	0
t_3	0	0	1	0
.	.	.	.	.
.	.	.	.	.
.	.	.	.	.
t_8	0	0	0	1

Therefore producing the input matrix (I) with also a corresponding output matrix(O):

$$I = \begin{pmatrix} 1 & 0 & 1 & 0 \\ 1 & 0 & 1 & 0 \\ 0 & 0 & 1 & 0 \\ 0 & 0 & 1 & 0 \\ 1 & 0 & 0 & 0 \\ 1 & 0 & 0 & 0 \\ 0 & 1 & 0 & 0 \\ 0 & 0 & 0 & 1 \end{pmatrix} \quad (9)$$

$$O = \begin{pmatrix} 1 & 0 & 0 & 1 \\ 1 & 0 & 0 & 1 \\ 1 & 0 & 0 & 0 \\ 1 & 0 & 0 & 0 \\ 0 & 0 & 0 & 1 \\ 0 & 0 & 0 & 1 \\ 0 & 0 & 1 & 0 \\ 0 & 1 & 0 & 0 \end{pmatrix} \quad (10)$$

Since the Petri net (figure 9) is directed with multiple sequential graph by edge connecting nodes. The column of a positive edge has a +1 in the row corresponding to one endpoint and a -1 in the row corresponding to other endpoint (Kirchhoff matrix properties). Then the degree of centrality is counted on the number of ties to nodes and out of nodes, which the numbers of loops are dependent. The loop dependence at the node provides the difference in potential of flow in which the incidence matrix (*IC*) (Output matrix − Input matrix) is generated. The incidence matrix shows the relationship between the places (n_i) and transitions (T). This same rule is used for all the other nets, which is a big advance with Petri modelling approach that one rule is defined and it applies for all cases.

$$IC = \begin{pmatrix} 0 & 0 & -1 & 1 \\ 0 & 0 & -1 & 1 \\ 1 & 0 & -1 & 0 \\ 1 & 0 & -1 & 0 \\ -1 & 0 & 0 & 1 \\ -1 & 0 & 0 & 1 \\ 0 & -1 & 1 & 0 \\ 0 & 1 & 0 & -1 \end{pmatrix} \tag{11}$$

On the other hand, for defining incentives based CFP that can lead to the motivation of multi-actor for participation in strategies for development and maintenance of ES, the argument is on pareto optimality. Lets the function g(x) denote the amount of contribution from actors Y (Local institutions, local investors, local government and banks) required to interact in balancing market for flow of x units of ES. Assuming that $g'(x) > 0$ and $g''(x) > 0$ for all $x \geq 0$, then g(x) is the real cost of contributing to maintain x units of ES of X. This is considered as the marginal cost is positive and increasing (Campbell, 1995). Similarly if H (x) denote the amount of contribution from another actors (stakeholders implying consumers, developers, migrants, suppliers, producers and markets managers, which are the multi-actors directly involved with the demand and supply of ES) and assuming that $H_i'(x) > 0$ and $H_i'(x) \leq 0$ for all $x \geq 0$, this means the marginal incentive for contributing in maintaining ES is always positive. But it is (weakly) decreasing leading to the characterisation of the set of pareto optimal allocations. It is always possible to find at least one pareto optimal allocation by maximising the sum of individual utilities subject to contribution/benefit to the resource constraint. Where u is the utility function of individuals (i = 1, 2, ..., m) and, H_i and Y_i are utilities of each individuals, then $\Sigma\ u_i = \Sigma\ H_i(x) + \Sigma\ Y_i$ and the solution to satisfy $\Sigma\ Y_i \leq \Sigma\ \Omega_i - g(x)$. Therefore, pareto optimality obviously implies that this inequality will be satisfied as equality (Feldman, 1980). This means in this situation that $\Sigma\ Y_i = \Sigma\ \Omega_i - g(x)$. Then considering a necessary condition for a maximum is $f'(x) = \Sigma\ H_i'(x) - g'(x)$, which is a function of x. It is also sufficient because $f''(x) = \Sigma\ H_i''(x) - g''(x) < 0$ for all $x > 0$ that means $f'(x) = 0$ implies $\Sigma\ H_i'(x) = g'(x)$, which is known as the Samuelson efficiency condition (Samuelson, 1954). Let x^* denote the value of x satisfying the Samuelson efficiency condition. Then, x^* is unique because $f''(x) < 0$ for all $x \geq 0$. The fact that the second derivative is negative means that f' is strictly decreasing as x increases because $f'(x^*) = 0$, then it must has $f'(x) < 0$ for all $x > x^*$. If $x < x^*$ implies $f'(x) > f'(x^*)$ and hence $f'(x) > 0$ and if $x = x^*$ and $\Sigma Y_i \leq \Sigma \Omega_i - g(x^*)$, then Σu_i is maximised subjected to the resource constraint and thus the allocation is pareto optimal. This means any allocation that provides exactly x^* units for contributing in a portfolio and uses exactly $g(x^*)$ unit of Y for the contribution and redistributes the remaining amount of Y to all stakeholders is pareto optimal.

Nevertheless, the challenge is determining growth model for the future condition that is uncertainty leading to the basic stochastic model that has the production function $y_t = \varepsilon_t f(K_t)$ (Ljungqvist and Sargent, 2000). K is capital contribution to the production function and ε is a random capital to the production function that depends on disturbances (interest rate, uncertainty of activities and others), while t is the time period. In any time period of consideration, t is divided into consumption C_t and the capital contribution going into the next period K_{t+1}. The state variables are k_t and ε_t, the control variables are K_t is chosen to give the realised values of the state variables, allowing the destruction of the condition of capital contributed after the defined period with certainty. The feasibility constraints are $C_t + K_{t+1} \leq \varepsilon_t f(K_t)$, $C_t \geq 0$, $K_{t+1} \geq 0$ and if this condition holds, then $K_{t+1} = \varepsilon_t f(K_t) - C_t$ as the transition equation. Therefore, there is no uncertainty about the next period of capital contribution with the same conditions (need for definitions). But there is certainty in capital contribution after a certain defined period say D because of uncertainty of state ε, which can be defined as $D_{t+1} = \varepsilon_{t+1} f(K_{t+1})$.

However, the problem is that of optimisation (Christensen and Kiefer, 2009), to choose the consumption function $C(K,\varepsilon)$, so as to maximise $V(K, \varepsilon) = E \Sigma^{\infty}_{t=0} \beta^t u(C_t)$. Where V is another consumption level and β is the disturbance, where the discount factor is $\beta \varepsilon [0, 1>$. The random variable ε must also be restricted. This makes it possible to characterise optimal consumption and saving behaviour and to analyse steady state for improving the performance in action for developing ES in term of parameters for contributing to their development. This follows the same rule of transition defined above. The management scheme controls the market behaviour following same normal rules of definition in the above situation for accepting and rejecting tokens. The behaviour of a net (with respect to a marking) is defined by a single rule that covers all the cases.

3.1.4 Execution and Implementation of Petri Nets Using Snoopy Software

A Petri net is executed as a mathematical graph that strictly defined syntax represented with appropriate tools such as Snoopy. The software package Snoopy 2.0 is used for executing and implementing the Petri nets definitions. It was used also for constructing the Petri nets and represents the graphic schemes into executable set of equations that the program can run simulations on them. Snoopy is a generic designed that facilities the implementation of different types of Petri net graphs (Rohr et al., 2010). The algorithm of its stochastic processes is built upon Gillespie (Gillespie, 1977 and 1992). This is an exact method that does a step-by-step simulation of possible states of the stochastic Petri net. It calculates the firing rates of all possible enabled transitions using their rate functions and combined the sum. The Petri nets were constructed and defined as a stochastic coloured Petri net (section 3.1.2) requiring the definition of colour sets, variables, functions and parameters that enable the performance of simulation.

Snoopy has a graphical editor for constructing Petri nets and its syntax is governed by the Backus-Normal-Form (BNF) formalism. BNF formalism is a special code for defining statements, assignments and expression, relational values, operators and parameters (Backus, 1959). The various colour sets defined for estimating the flow of ES are implemented in Snoopy using this formalism based on the kinetic specifications (see section 3.1.3). In table 6 colour sets defined in the software for execution is presented. They are different colour categories such as colours for flow of ES, finance, flow of management schemes (see section 3.1.1) and combined colour sets as "product" and "union".

Table 6: Definition of Colour Sets for Implementing Petri Nets

Name	Type	Colours
Dot	dot	dot
Compset	enum	C_1, C_2, C_3
ES	enum	$ES_1, ES_2, ES_3, ES_4, ES_5$
Fin	enum	a,b
MS	enum	MS_1, MS_2, MS_3
PP	Product	Compset, ES
Market	Union	PP, MS, Fin
Management	Union	PP, MS

Dot is a colour set category definition, which allows the other colour sets to be defined under it. Colour set categories of components of ES can then be defined under Dot. Compset is defined in Dot as the colour set of markings of components of ES containing $Comp_1$, $Comp_2$ and $Comp3$ represented in the colour formalism as C1, C2 and C3. This follows the same rule for all the definitions of colour sets and categories that were established in section 3.1.1. ES is a set of tokens defined by different colours such as ES1, ES2, ES3, ES4 and ES5. Fin is a colour set for financial flow defined with colours "a" (financial benefits) and "b" (financial contributions). MS is a colour set for the marking management scheme with colours defined as MS1 (permits), MS2 (certification) and MS3 (conservation credit/banking). PP is a product of colour sets compset and ES (Compset, ES). Market is a union for colour sets PP, MS and Fin (PP ∪ MS ∪ Fin). Management is a union of the colour sets PP and MS (PP ∪ MS). All these markings as colour sets containing colours are tokens that flow through the net, but only on places that contain the definition of the type of colour set category. In a place of the Petri net the colours can be assigned as single or aggregates of all colour sets categories. For example, the colour set PP can be assigned as single colours entry in the marking list at a place:

$\{(C_1ES_1), (C_1ES_2), ..., (C_1ES_5)\}$
$\{(C_2ES_1), (C_2ES_2), ..., (C_2ES_5)\}$
$\{(C_3ES_1), (C_3ES_2), ..., (C_3ES_5)\}$

This means a transition can be fired to remove only a single colour or variables that are group of colours. On the other hand, aggregates of all the colours for ES can be implemented in the list of marking at a place as "all ()". It means an aggregate of all the ES types as one variable for the flow. This means all the single colours of this colour set. These rules apply for all colour sets and places in the net. After assigning the colours or its aggregate of colours, the numbers of tokens corresponding to each of the colour or aggregate in each of the places are entered. Variables are used for expressing the flow of tokens and their rate on the Petri net as single or combination of tokens types with regards to the colours or colour sets definition of places they are mapping (on the Petri net). The rates of flow of tokens or variables are expressed on arcs and they also determine the arc weight. The expressions control the kinetic specification of a transition that would be enabled for tokens to flow through the arc already defined (see section 3.1.3). The following are variables defined in the Petri net for colour sets formalism:

X   -   Compset
Y   -   ES
M   -   Fin
N   -   MS

However, if the flow expressions are not defined in a coloured Petri net, a simulation cannot be performed because no flow will be enabled when a transition fires. The expression may be single values, subset of a variable, combinations of variables or their subsets of variables or their inter combinations. They are expressed with respect to the code of the BNF syntax formalism. The following are some examples of expressions executed on the arcs:

a	-	Only a flows on the path of the arc
b	-	Only b flows on the path of the arc
m	-	Only m flows on the path of the arc (a or b)
n	-	Only n flows on the path of the arc (MS_1 or MS_2 or MS_3)
(x, y)	-	Only the colour set (x, y) flows $\{(C_1ES_1), (C_1ES_2), ...,(C_3ES_5)\}$
a ++ b	-	Only a and b flows
n ++ (x, y)	-	Only n and (x, y) flows
$[x = C_1 \& y = ES_1]$ (x, y)	-	Only the flow (C_1, ES_1)
$[x = C_3 \& y = ES_2]$ (x, y)	-	Only the flow (C_3, ES_1)
$[x = C_1]$ (x, y)	-	Only the flows contain C_1 that is $\{(C_1ES_1), (C_1ES_2), ...,(C_1ES_5)\}$
$[n = MS_1]$ n	-	Only the flows MS_1
$[n = MS_2]$ n ++ (x, y)	-	Only the flows MS_2 and (x, y)

They are examples of expressions to specify the flow of variables or other entities according to the BNF code. Figure 12 presents a defined Petri net for modelling multi-agent behaviour for estimating ES. It was established before (see figure 9), but defined here by introducing in the Petri net the implementation and execution definitions. It shows that each place on the net contains 15 tokens, which represents the ES defined in the colour set PP (aggregated colour set definition). The places "MES", "Demand unit for ES", "Supply for ES" and "Stock of ES" contain 15 tokens each as the colour set PP. For any transition to fire to remove PP from it input to output places, both places must contain the PP colour definition. If the transition producers fires it removes PP (15 tokens) from MES (input place) and put it to Supply unit for ES (output) and so on for all other transitions. This is possible because they have the token definition at each place. The (x, y) on the arc means only the PP defined tokens flow. A management scheme can be used for balancing ES (improvement of deficit or upsetting surplus) through MES. This means surpluses from estimation of balances of ES at the landscape can be upset in this MES framework, while deficits can be mitigated. Therefore, for non degraded landscapes the management scheme can be use for upsetting surplus due to environmental improvement. While on a degraded landscape the MES for improving ES due environmental deficit.

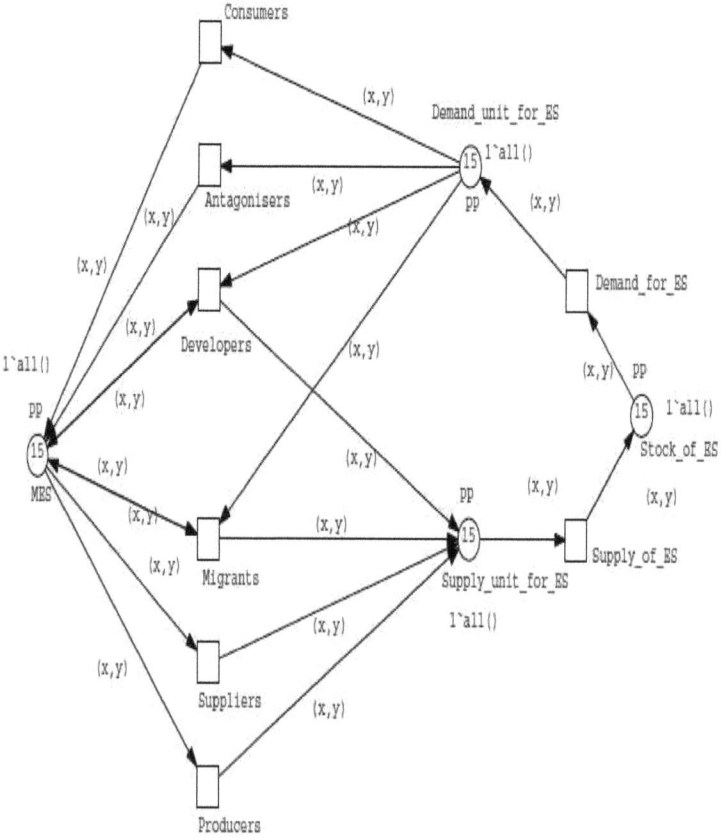

Figure 12: Defined Petri Net for Modelling Multi-agent Activities for Estimating Ecosystem Services

Figure 13 presents a defined net of management scheme for balancing ES within internal and external markets for exchange of MS. It was also established before (see figure 10), but defined here by introducing in the Petri net the implementation and execution definitions. It shows that each place on the net contains 3 tokens, which represents the management strategies defined in the colour set MS (aggregated colour set definition). The places "Management scheme", "Market scheme" and "External market scheme" contain 3 tokens each as the colour set MS.

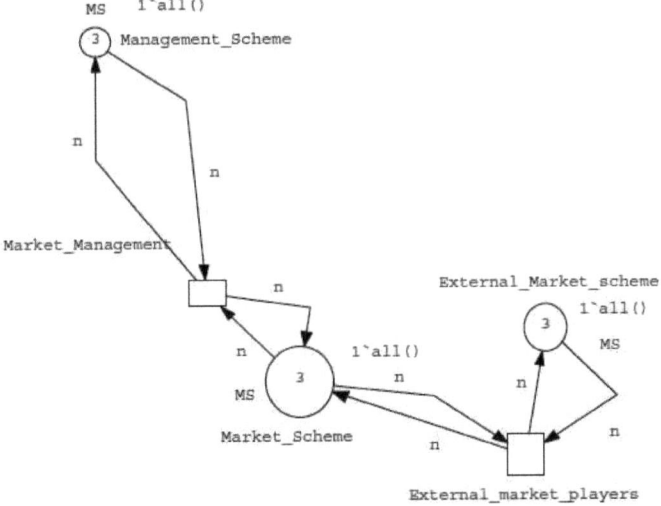

Figure 13:   A Defined Petri Net for Modelling Management Scheme for Market for Ecosystem Services

If the transition "Market management" fires, it removes MS (3 tokens) either from the place "Management scheme" or Market scheme". When from "Market scheme" (input place), then it put to "Management scheme" (output) and vice versa. The n on the arc means only the MS defined tokens flow. To encourage these markets that would support the preservation of ES to grow, incentive and motivation schemes as CFP are used to encourage markets that would encourage the preservation of MES. A formal introduction of the relationship of CFP was established in section 3.1.3. Figure 14 presents a defined Petri net for modelling CFP for investment in MES. It was also established before (see figure 11), but defined here by introducing in the Petri net the implementation and execution definitions. It shows that each place on the net contains 2 tokens, which represents finance defined in the colour set Fin (aggregated colour set definition). The places "Financing", "Contribution unit", "Financial portfolio" and "Benefits unit" contain 2 tokens each. If the transition "Flow of contributions" fires, it removes Fin (2 tokens) from the place "Contribution unit" (input place) to "Financial portfolio" (output). The expression a, b, a ++ b on the arc means only the flow of Fin defined tokens flow as different aggregates of them. That is financing with the flow a (financial benefits) and b (financial contributions), m (a or b) and a ++ b (a and b).

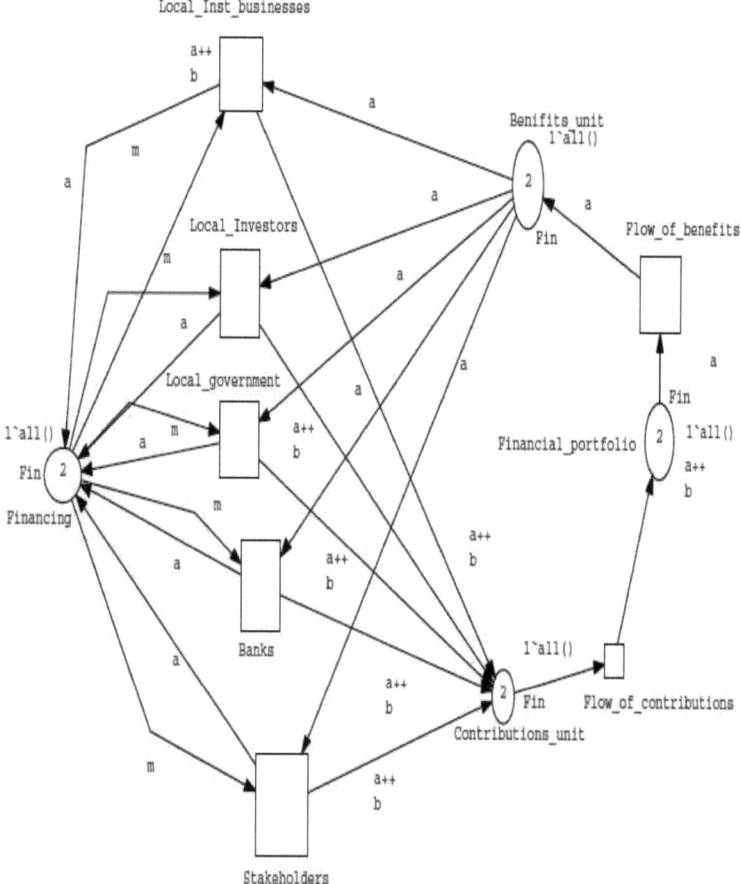

Figure 14: Defined Petri Net for Modelling Community-based Financial Participation for Developing Market for Ecosystem Services

Where Local_Inst_businesses defined in the figure stands for local institutes/businesses. A unified system for business modelling for developing markets to preserve ES is presented by coupling the 3 modelling frameworks defined in this section (Krishnamurthy, 1989). Putting the models together can provide a management framework for financing, estimating and balancing ES at the landscape through markets (MES and MS). Figure 15 presents a defined unified Petri nets for business modelling for markets to preserve ES.

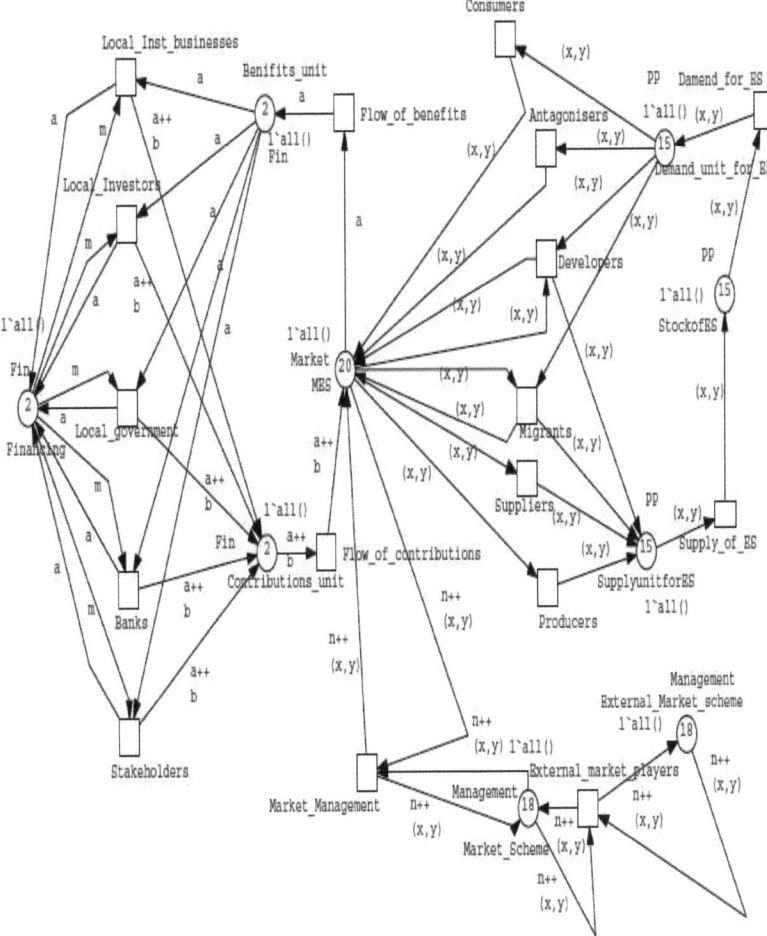

Figure 15: Defined Unified Petri Net for Business Modelling of Market for Preserving Ecosystem Services

They are the 3 modelling frameworks defined Petri nets (see figures 12, 13 and 14) that has been coupled at their interfaces MES, Management and financial portfolio to provide a place "Market". This is because financial portfolio is required to develop MES. The arcs connecting MES, Management and financial portfolio are decomposed and recomposed to the place "Market". The colour set and variables at that place "Market" is specified to also couple all the colour sets definitions of the 3 Petri nets places that were brought together. That is the colour set Market for PP, Fin and MS. Then a union set is used to integrate them as unified flow systems (Market). It shows that at the place "MES" there is 20 tokens comprising of colour definition of the 3 nets corresponding to 15 from place "MES", 3 from "Management scheme" and 2 from the "financial portfolio". On the other hand, the management scheme places contain 18 tokens as colours PP and MS corresponding to the union of the colour set Management. The tokens MS has to flow with the tokens of the colour set PP (x, y) for balancing the market. While the flow of "fin" in the CFP net and PP in the net of multi-agents remains the same colour sets as before. The models run depending on their functions or guard values (established as Boolean or switch in section 3.1.3).

A function can be assigned as "TRUE" for enable and "False" for not enable. A guard evaluates to either true or false for the flow of tokens. A token can only flow over arcs, when the guard evaluates to true. The expression will typically require the case attributes that have to be assigned to a function or guard based on the data or other information. A stochastic simulation is performed using "Gillespie" simulator in Snoopy with 100 output step counts to generate the results of the behaviour of the input data (see section 4.3.2). There are many simulators in Snoopy, but Gillespie is used because it is well known to generate statistically correct trajectory (possible solution of stochastic equations) (Gillespie, 1992). The nets are tested and the behaviour of each of the study variables simulated and they are checked and corrected to generate the same results as the unified net. This can allow the verification and validation of the nets (see section 3.1.5). They are checked with different random values to examine that they are not a static model and also whether each of the nets can simulate the same results as the unified net. This was to insure that the net could be able to simulate the data from UNESCO BRS, which the simulations are established in sections 4.3.2 and 4.4 with the aggregated data system in section 4.3. This was guaranteed, which then leads to the verification and validation of Petri net properties.

3.1.5 Verification and Validation of Properties of the Petri Nets

The verification and validation of Petri nets are analysis that can help increase the accuracy of the model (Chiola et al., 1989; Castanet et al., 1985; Sloane and Gelhot, 2004). In table 7 a summary of Petri nets properties that are analysed for verifying and validating the net constructions (see figures 12, 13, 14 and 15) is presented. It shows that all the structural properties of the nets are correct (bounded, conservativeness, repetitiveness and consistency). The general problem was the topology of the net, which its liveness can not be decided. A deadlock test and reachability analysis was carried by performing a slow simulation for all the nets. No state was noticed that a transition cannot fire or run into deadlock state. This was confirmed again by performing simulations of 100 run output step counts and 1000 run again, which did not detected any state of deadlock. The liveness property of Petri nets is the detection whether all the tokens can reach all the places as defined by the net. For large constructed Petri nets, it is sometime difficult to determine the liveness due to an extended wide reachability graph (a sketch of the topological structure). This can be proved with the theorem of extended free choice Petri nets (Desel, 1992).

Table 7: Summary of Verification of Petri Nets Properties

Net Properties	Multi-agent net	Management Scheme net	CFP net	Unified net
Net class	Ordinary	Asymmetric choice	Extended free choice	Ordinary
Liveness	Cannot decide	Cannot decide	Cannot decide	Cannot decide
Boundedness	Yes	Yes	Yes	Yes
Conservativeness	Yes	Yes	Yes	Yes
Repetitiveness	Yes	Yes	Yes	Yes
Consistency	Yes	Yes	Yes	Yes

Boundedness is a test to examine whether the flow of tokens on the Petri net can reached a system state as required. Conservativeness is the detection test of the number of tokens that can re-site in each place of the Petri net. Repetitiveness examines whether a system can be repeatable. That is continuously firing, if the net do not stop, the property is satisfied. Consistency determines the frequency of a transition firing, which need to be consistent with the sequential path of the flow of tokens through the nets. Real scenario data are then introduced into the Petri net (see figure 12) to model the resulting behaviour. Therefore a data sampling strategy was established for data collection and processing for input data for real scenario application of the Petri net.

3.2 Data Sampling Strategy

A good understanding of data sampling and why they are used is the central to designing a credible decision support framework. The sampling for data collection is based on a cross-sectional field observation at the UNESCO BRS (see section 4.2). The UNESCO BRS is located in Brandenburg and is taken as an experimental area for testing the concepts that are being established within theoretical background (see section 2) and the modelling framework (see section 3.1). It is also the first attempt of experimenting concepts of ES in this region. This is to estimate the capacity of ES at the landscape scale, which is used as input data for the modelling framework (see section 4.4). The data source comprises of landscape components that were established in section 2.1, which can support the estimation of ES. They are based on activities that lead to their balance, improvement and deficit (see table 2), their classes and examples (see table 1) including indicators of agro-forest ES (see table 3). They were estimated by qualitative value judgement based on ranking them to a scale of 0 to 5 based on an observation survey. This was done from 02 August 2010 to 31 August 2010 and a few field re-observations for data quality control in September and October 2010. Burkhard et al. (2009) had also used this ranking scale to map ES on different scales. This ranking system is used to translate qualitative observatory data to quantitative data that can allow scenario analysis for management support systems. The ranking scale is defined as the followings:

- 0   no relevant capacity
- 1   - very low relevant capacity
- 2   - low relevant capacity
- 3   - medium relevant capacity
- 4   - high relevant capacity
- 5   - very high relevant capacity

The EC are also using a ranking scale of 0 to 3 for capacity assessment of institutions (EC, 2004). Data collection based on scale ranking is very common for primary data (original data) sampling and have been used by many researchers for estimating different variables (Adèr et al., 2008). It has a clear advantage that it is not time consuming as compare to taking real measurements. It is also suitable for short term research projects and it is cheap. But the problem with such a method of data collected is that a lot of data quality management have to be put in place (Batini and Scannpieco, 2006; George et al., 2000; Adam and Wortmann, 1989; Pipino et al., 2002; Winkler, 2004). Therefore Bisgaard (2008) and Pearl (2000) argued that for such type of data collection method, quality management of samples must be the most important aspect of it and need to be put in place before conducting it.

3.2.1 Preparatory Set Up for Data Collection

Preparations started by designing a sampling strategy as criteria for quality data collection. The first step was identification of the sampling region in terms of administrative units that belong to the UNESCO BRS. Some administrative maps of the Lausitz region, which show counties was collected from the state of Brandenburg Cartography and Measurement office and also maps of the UNESCO BRS from the authority of the BR. They were used to identify the administrative units and a Geographical Information Software (ARCGIS) was used to establish a map of UNESCO BRS within the 3 counties. That is Upper Spreewald-Lausitz, Spree-Neiße and Dahme-Spreewald. An internet satellite map was used as a tool for identification of particular areas by zooming onto one spot to another with the lowest distance capture of 200m (see http://www.citypopulation.de/php/germany-brandenburg.php). It shows streets, roads, railways and even the boundaries of the UNESCO BRS, but has no information on landscape components. This was to mark out specific area of consideration for data collection (sampling areas), which these areas and their surroundings were the main areas of data collection. The following main sampling areas were chosen under each of the counties:

- Upper Spreewald-Lausitz: Lübbenau
- Spree- Neiße: Burg
- Dahme- Spreewald: Lübben, Straupitz, Schlepzig, Krausnick, Märkische Heide

This was then followed by field feasible studies to estimate the accuracy of the satellite map and the online measurement tool that was to be used. It was seen as reliable. During the feasibility study, it was noticed that the whole region of the UNESCO BRS is labelled with road/route signs leading from one area to another with specified number of kilometers that was seen as an advantage to ease the data collection process. But some of the kilometers counts were within different areas of consideration and also not in km^2. Therefore an online tool for land measurement from the Brandenburg Cartography and Measurement office was used to specify kilometers and boundary areas to avoid overlapping within areas or inclusion of one region that have already been considered in another observation (see http://geoportal.lkspn.de). Sampling worksheets where prepared for entering data that comprise of two types of matrix tables. All of them have columns of the main 5 categories of ES with one having a row of the main landscape components (land surface, biodiversity and water).

The other comprise of rows of components for MES (carbon sequestration, biodiversity and water), actors for MES (consumers, suppliers, developers, producers, antagonisers and migrants) and service units (demand and supply). The two matrix tables with some already existing maps of the UNESCO BRS from administrative offices (the authority of the BR) became the working sheets for entering data. Maps from tourist information in Radusch were used for route accessibility of specific areas within the field, especially as some areas where having thick forest that require a guiding map to penetrate them. A bicycle was used as a mean of transport from one located area of observation survey to another. Areas that where not accessible using a bike, a paddling boat was used. The internet satellite map was used again to prepare a plan of the areas that have to be covered for the next day after each day of sampling. This was to avoid the repeatability of areas that were already observed in the samples. The data collection scheme (see section 4.2) and analysis of their results (see section 4.3) are presented in section 4.

3.2.2 Data Aggregation Procedures

The samples were collected in areas of different km^2, which cannot be aggregated without normalisation because areas with higher or lower km^2 may affect the weighting of the samples. Therefore they were brought to a common km^2 before aggregation. Normalisation is a means of allowing data with different scales to be composed or compared by bringing them to a common scale in order to remove statistical errors (Von Eye, 2003; Eelko, 2007). But one can also argue that for representativeness of data that are collected at different range of distances, they also need to be brought to a common distance before working with them to avoid the effect of weighting rates. This can be justified from the theory of preserving homeomorphism groups to satisfy uniqueness (Luce, 1986; Alper, 1987). Therefore the samples were aggregated by multiplying their value with their areas and dividing them with the sum of all the areas. That is;

$$\sum_{i=1}^{m} (n_1 x_1 + n_2 x_2 + \ldots n_m x_m) / \sum_{i=1}^{m} (n_1 + n_2 + \ldots n_m)$$

where n is the number of km^2 and x is the variable of the sample (ES). To aggregate the data for the designated areas under the administrative unit of Straupitz, their areas in km^2 were considered for normalisation of the data. In table 8 an estimate of the sampling points in the administrative unit of Straupitz in km^2 is presented.

Table 8: Sampling Units at the Administrative Unit of Straupitz

Sampling units	Estimated Areas in km²
Straupitz central to North Straupitz	4
Straupitz to Muhlendorf	9
Straupitz central	4
Byhleguhre to Straupitz	24
Straupitz to Neu Zauche	12
Neu Zauche to Wußwerk	16
Wußwerk, Burglehn to Alt Zauche	28
Alt Zauche to Bukoita	12
Burg to Byhleguhre	16

The rank value for ES that was given to each of these areas was multiplied by their km² and sum up for all the areas and divided by the sum of all their km². This removes the negative effects of the samples and also brings them to a common unit. After doing the first normalisation, then the samples can be aggregated or decompose from one level to another without further normalisation. They were further aggregated from one level to the other by simply looking for their mean values. Since the interest was to estimate them, there was no correlation or looking at their sum of squares or determinants. These mean values then became the input values to the model that were simulated for decision support systems for balancing ES at the landscape scale. It is difficult to apply decimal numbers in the software tool. So the mean values were all converted to a scale of 100 (percentage), which removed all the decimals before encoding them on the Petri net (see section 3.2.3). The nets were defined to have aggregated the types of ES as 15 classes requiring that all input data are also encoded with respect to this structure. In order to interpret data analysis and simulation results also typical estimation parameter of the value of ES were needed, which was established (see section 3.3) and use for interpretations of results (see sections 4.3 and 4.4).

3.2.3 Quality Assurance of Data

A quality control is put in place on the resulting sampling data estimated at the UNESCO BRS. The sampled data were re-checked, especially areas that call for a discussion by a supported re-examination and field re-checking for correctness. All these procedures were to ensure that datasets satisfied some degree of quality assurance before encoding them in the modelling framework for simulation. After checking the data for their correctness, they were converted to another scale that was easy to work with while sustaining the ranking scheme. This was to eliminate decimal numbers and bring it to a form that can ease the application with the software tool. In table 9 a conversion scheme of the ranking scale is presented.

Table 9: Conversion of Ranking Scale

Ranking Scale	Conversion % (1500)	Meaning
0	0	No relevant capacity
1	20 (300)	Very Low relevant capacity
2	40 (600)	Low relevant capacity
3	60 (900)	Medium relevant capacity
4	80 (1200)	High relevant capacity
5	100 (1500)	Very high relevant capacity

It is converted to a percentage and shows the various level of ranking corresponding to the formal one. Then other scales can also be converted in the same way, while preserving the ranks, which can be multiple or division. Multiplication or division of numbers can be converted to this ranking scale. One can easily depict its corresponding rank within the scale. For example multiplication of a variable by 15 that have been converted to percentage will give 1500 and their corresponding ranks are 0, 300, 600, 900, 1200 and 1500 respectively. This can support the interpretation of parameter estimates.

3.3 Parameters for Estimating the Values of Ecosystem Services

The value derived from estimates requires typical parameter that can guide their interpretability. The following parameters are assumed:

$$ES = TES + NTES \tag{9}$$

$$VES = \Sigma (ES \pm Adj. \pm (SPU - SAU) \pm \beta) \text{ or } \Sigma (ES \pm Adj. \pm SU \pm \beta) \tag{10}$$

$$St = \Sigma (VES \pm (SS - DD)) \tag{11}$$

$$\Sigma VES \pm (SS - DD) = \Sigma VES \pm \Delta\Sigma VES \tag{12}$$

Therefore substitution of their values will form the following:

$$\Delta St = \Delta\Sigma VES \pm (SPU - SAU) \pm \beta \text{ or } (SS - DD) \pm (SPU - SAU) \pm \beta \tag{13}$$

If a certain St is assumed or had been estimated at a particular place before a transition event, then it is considered as the old stock of ES due it a reduction or increase in some services. This will lead to a change in stock. The total of stock at a particular region or place is the sum of all the stock before and the sum of the different after a transition (current stock):

ΣSt - Old stock of ES (before a phase transition) in a particular region

$\Sigma St \pm \Delta\Sigma St$ - Current stock of ES (after a phase transition) in a particular region

For environmental balancing strategy for preserving ES:

$\Delta\Sigma St = 0$ or $\Delta\Sigma St \geq 0$ (14a)

For environmental improvement strategy for preserving ES:

$\Delta\Sigma St > 0$ (14b)

For environmental deficit strategy for preserving ES:

$\Delta\Sigma St < 0$ (14c)

Therefore, if a permit is assumed as a market strategy for environmental balancing to preserve ES, then it can lead to value creation for business based on the price of acquiring or selling permits. This will require the role of market forces (demand and supply that determines the price). Environment balancing measures require that $\Delta\Sigma St = 0$ in the case where demand and supply of ES are balance within a regulatory target or threshold level or above. While on the other hand, in the case of balancing strategy that the target values are below a threshold or regulatory target, then $\Delta\Sigma St \geq 0$. This confirms to the laws of limits in calculus and can be applicable in economics and other fields where there are upper and lower bounds for estimating parameters (Freeman, 1995). But this needs to be supported by regulatory measures to encourage payment for damaging the environment or ecosystems. These concepts can be used to derive strategies to support the preservation of ES for nature protection. Environmental improvement ($\Delta\Sigma St > 0$) means that permits can be offset in the external market and their cost would reduce in the internal market (market conditions due to demand and supply) leading to increase in social welfare (value creation for preserving ES). On the other hand, for environmental deficit for preserving ES ($\Delta\Sigma St < 0$), permits would be acquire in the external market and their cost would increase leading to a decrease in social welfare. In this type of scenario, if incentives are not given to develop ES, then it can possibly lead to a scenario of environmental degradation. Therefore ΔSt offers the potential for market development, which need to be developed, especially in protected landscape to fight land degradation and desertification. This means the area of focus for business development for preserving ES is $\Delta St = (SS - DD) \pm (SPU - SAU) \pm ß$.

4. Model Application to UNESCO Biosphere Reserve Spreewald

The Petri net modelling framework and the data sampling strategy established in the methodology (see section 3) are applied at the UNESCO BRS. UNESCO BRS is located about 100 km to the south east of Berlin in state of Brandenburg (Germany) in the Lusatian region. It is a wetland of about 475 km^2 surface areas under three main administrative districts (Dahme-Spreewald, Spree-Neiße and Upper Spreewald-Lausitz) with about 50,000 people of the Slavic tribes (Sorbs/Wends) and others who migrated there. The BR in Spreewald was established by the UNESCO administration as a protected area to preserve threatened species of biodiversity and conserve the degraded landscape (Natura, 2000; Vahrson et al., 2000; Jenssen and Hofmann, 2004). It was created in Spreewald with regards to the combinations of the regulation § 6.1 of the German Environmental Law of 29 Juni 1990 and §§ 12, 13 and 15 of the German Nature Protection Law that applied to the region to became a protected area (Bronner et al., 1997). In 07 March 1991, the region was recognised as a protected area by the UNESCO administration. Figure 16 presents a map of the UNESCO BRS with the main data collection areas and the different zones of protection with their different status. It shows zones of different level of protection with separate colours. The zones are protected under the following status:

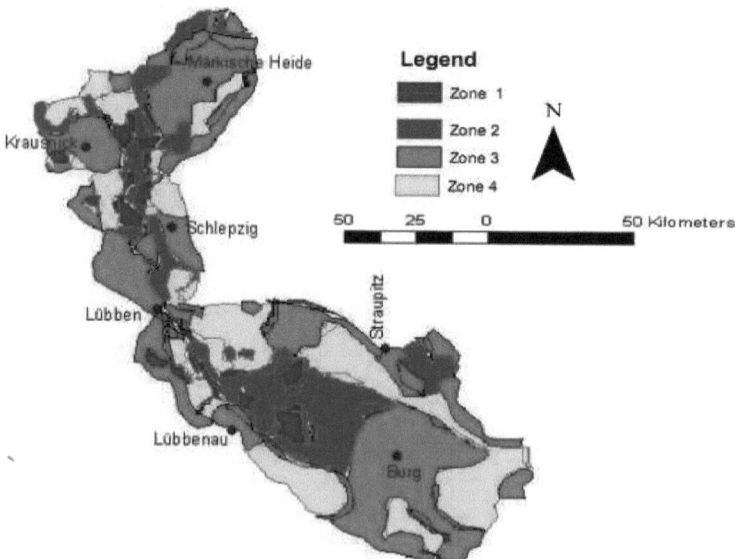

Figure 16: Map of UNESCO Biosphere Reserve Spreewald with Zones of Protection

- Zones 1 are the core nature protection landscape with high protection status that fully remains under natural dynamic for research and studies (2% of the whole BR). Especially in the lower part of the administrative units of Lübbenau and Burg
- Zones 2 are isolated landscapes due to degradation, reserved for protection of green plants and animals species (18% of the whole BR)
- Zones 3 are development and agricultural protected areas. There are combined nature and culture protection areas. It has aesthetic and cultural values, which has attracted a lot of tourist activities in these areas. It has a limited protection status (43% of the whole BR), and
- Zones 4 are regeneration landscapes due to degradation with some development and agro-forestry activities (37% of the whole BR)

The landscape type is mainly agro-forestry with many water systems of canals, streams, lakes, ponds, rivers and swamps with a few settlement areas for the population that have preserved their culture. These combined with the biodiversity of the region to influence the growth of the tourism industry with frequent boat trips, bike tours and sporting events. The settlement in this region developed as a result of some main projects between 1907 and 1974 such as the following:

- Construction of road track into the Burg region that encourage settlement of people
- Construction of streams and canal systems in inner Spreewald that paved the way to the canoe and boat tradition in the region
- Development of Lausitzer brown coal mining for energy production that created employment, and
- Use of water for commercial agriculture

At the end of Berlin-Görlitz railway in 1866, tourists were being attracted in the region leading to the creation of the union of boat paddlers (Succow, 1992). By 1989 about 1.5 to 1.8 million tourists came to the region that were being attracted by boat trips and boat paddling, especially in areas like Lübbenau, Leipe, Wotschofska and Lübben waterways. Then by 1992 the number of tourist increased to about 2.4 million and it is still increasing till today with those areas still being the major attraction. The main economic activities of the local population are tourism, agriculture and some few tertiary activities with the development of the cucumber, millet, and vegetable tradition, herbs for cosmetics purposes, forestry, fishing and hunting (currently highly restricted). But many farmers are now diverting to energy plants due to the high market prices for them. The landscape is turning to agro-forest with continuous reduction of forest and green land influenced by extension of agricultural land. Nair (1989) defined agro-forest systems as the deliberate growth of woody perennials on the same units of land used for agricultural crop and/or animals, with significant environmental or ecological interaction between the woody and non-woody components of the system. Farmers or pastoralists plant trees on their land to fix nitrogen and provide shade that also serve as source of fruit or nut, or for additional expected income from timber sales. Agro-forest system can also be the reduction of forest land for agriculture with a mixture of forest and farming system in the same parcel of land.

Furthermore, many restaurants, hotels, holiday houses, leisure parks and other tourist facility have been constructed in the region that have influence the continuous growth of tourism. This has also an impact on the landscape, which are negative if not controlled. This section is structured with section 4.1 presenting an overview of the landscape of UNESCO BRS and section 4.2 the data collection and application of aggregation system established in the methodology (see section 3.2.2). Section 4.3 presents the results of estimates of ES at the BR, while section 4.4 presents modelling and simulation results for managing ES in the region. Section 4.5 presents modelling and simulation results for business development for preserving ES at the BR and section 4.6 established discussions of the results.

4.1 Landscape Identification

The land cover of Spreewald region is mostly agro-forest with forest, farmer land, greenland, water and other land cover types like the settlement areas for the population. The forest cover maps from 1751, 1846 and 1939 in Krausch (1960) shows that the region was first a highly forested landscape that have undergone a lot of land use changes. That is by 1751 the whole region was almost cover with forest and by 1846 a lot of forest land has disappeared in the upper Spreewald along Burg, Raddusch, Werben, Schmogrow and Byhleguhre, Straupitz to New Zauche and Regow to Krimnitz and part of Lübbenau were standing without forest. While by 1939, most of the region was without forest with small forest area between Byhleguhre and Straupitz, which is presently the worst forest area with death trees, soil fragmentation and lack of regeneration of both the soil and trees. The only big forest area within this period is in the area from Alt Zauche along Wotschofska to Burg. When comparing these maps with Corine land cover 2000 (see http://biodiversity-chm.eea.europa.eu/information/document/F1088156525/F1125582140) of the region show that the Upper UNESCO BRS had only a minor increase in forest land cover. Especially areas around Burg, Werben, Schmogrow and Byhleguhre were still standing without forest, while some improvements of forest cover in the other regions. The Under UNESCO BRS region is not mention from 1951 to 2000 due to lack of data. But the Corine land cover 2000 map shows that the area is highly forested with development of swamps in its inner parts. This is the same situation in all the inner parts of the BR (mostly zone 1 and 2) with a small swampy area in Byhleguhre that is not in the inner part. The land use changes of one landscape component can be better understood by examination of it with other land use components. Figure 17 presents an analysis of land use components from historic time of UNESCO BRS in 1991 to 2010. It shows that from 1991 to 1999, farm land use and settlement/others have reduced, while forest land use was increasing till 2001 that then continue to decrease until 2010. On the one hand from 1999 to 2001, there was a sharp increased in farm land use and continue until 2010.

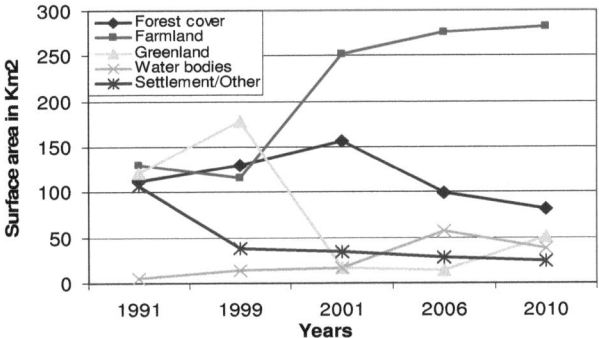

Figure 17: Land Use at the UNESCO Biosphere Reserve Spreewald from the Years 1991- 2010

The green land has dropped drastically within this period and remains constant from 2001 to 2006, but increased again from 2006 to 2010. On the other hand from 1999 on to 2010, the settlement/others land cover constantly reduced slightly, while land use for water has been gradually increasing from 1991 to 2001. Then by 2006 a sharp increase was noticed, which drop again in 2010. The data for 2010 was considered before the rains in August and the strong winds on the 16 July that cause a lot of disaster like bringing down trees and causing flooding in the region. This was also controlled not to have an influent in any data, but they were indicators for the need of sustainable strategies. The UNESCO BRS is crossed by the river Spree and more than 300 canals, which some are presently block as standing waters. These water networks contribute to a great biodiversity of flora and fauna in the region. Figure 18 presents a map of the water system at the UNESCO BRS, which shows that they flow from the Upper BR to the Under BR. They comprise of many type of water bodies that can be interpreted from the legend. The main water flows are from the north to south of the BR such as the Nordumfluterer, big streams (Mutnitza), Burg-Lübbenau-Canals, Main Spree and Südumflutter.

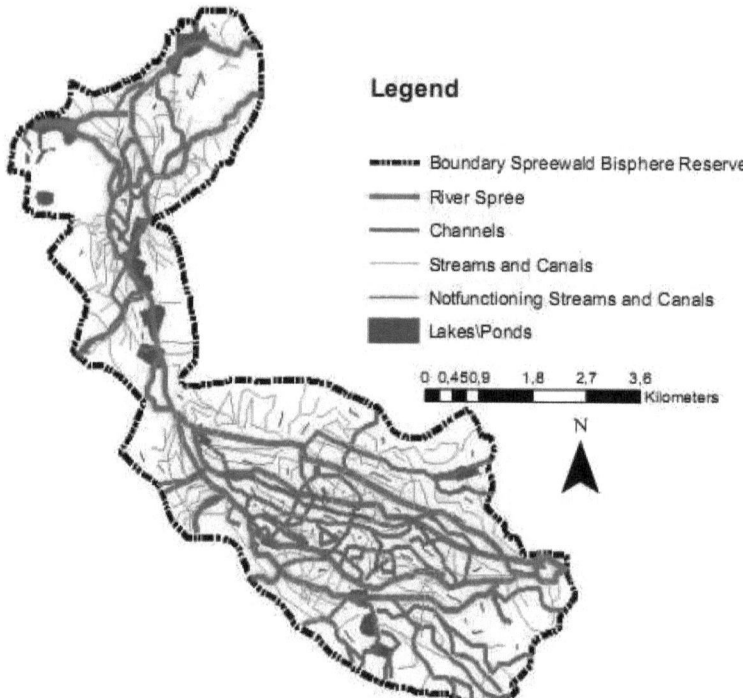

Figure 18: Water System of UNESCO Biosphere Reserve Spreewald

There are a lot of lakes like Heide, Köthener and Neuendorf mostly in the Under UNESCO BRS, and also ponds like in Börnichen, Petkamsberg and Fließdorf. The streams and rivers re-unit in Lübben waterbed then split again to the Wasserburger Spree, Puhlstrom, the river Zanias and main Spree. The Wasserburger Spree flows as a minor canal in the inner Under BR through Wasserburg in the direction of Heide and Köthen lakes. Then Dahme-Umflut canal and through the main Spree in the direction of Neuendorfer Lake heating the Dahme-

Umflut canal at the lower part of Leibsch. In the inner parts of BR (Zone 1 and 2) are swamps and also some ponds for fishing (Succow, 1992), which the situation is still the same till today. These ponds and swamps combined with the stream and rivers (braided streams and rivers) in contributing to organic matter in water that threatens the water quality. Even though along the streams and river organic material is being removed, but the standing waters and swamp produce macroprophytes contributing to eutrophication conditions. Leafs of trees, fertilisers and chemicals (like the use of fungicide and herbicides in cucumber farms in Stradow) are washed from farms into these water bodies. Flood plains in cultivated landscape also exchange nutrients that fuse into the water system combining leafs and oxygen to provide favourable conditions for phytoplankton growth. These organic substances contribute to oxygen through nitrification giving room for the growth of algae boom in water (Calow and Peet, 1994; Allan, 1993), thereby threatening the water quality. There is also high humus contain on the soil surface with usually an acidic pH value and nitrogen of 0.1% to 0.3%.

The soil conditions also have a low value of phosphor and calcium that are the main nutrients in soil (MLUR, 2000). Presently the soil condition in areas around the Byhleguhre and at the Krausnicker Berg (from Krausnick up slope) are shrinking and being fragmented causing landslides, which trees are uprooted. The brawn colours of the trees are also indicating dead soil with little water retention capacity. But within the valleys where there are lakes, swamps and water flow the trees are green due to availability of water in soil. Therefore the roles of water resource are very essential for landscape management for preserving ES. The land cover type and its component also influence the biotope in the region, which contain flora and fauna. The availability of flora and fauna in the region reflects the land cover type like areas with large forest cover has abundance of species such as fox, squirrel, mouse, forest herbs, grasshoppers, coniferous trees, pines and many others. They are mostly found within zone 3 and 4 of the BR. Swampy forest and grassland species such as reptiles like frogs, lizards and amphibians are mostly found within zone 1 and 2 at the BR. The water species are fish, crabs and colour birds around water bodies. While farmland species are crops like cereal, corn, vegetables, animals like cattle, sheep, pig, goats and chickens are found mostly within zone 3 and 4. The settlement areas are within zone 3 that contain gardens with vegetables and some insects, ants and garden animals. But the UNESCO BRS does not have any wildlife. According to Jentsch and Klaeberg (1992), the last wild animals like wolf and wild cat disappeared in the Spreewald region in the year 1844. The intensity of landscape components such as biotopes, biotic and abiotic factors are spread with respect to the zones of protections. Petschick (2010) provide a biotope map of UNESCO BRS with different classes of abiotic and biotic species. The greatest amounts of biodiversity are in forest and greenland swamps within zone 1 and 2. While farmlands, settlements and other are being concentrated within zone 3 and 4. The greatest portion of greenlands are pasture lands and are found in moist soil, wet or flooded areas (zone 2 and 3). These characteristics of the landscape determine the biotopes in the UNESCO BRS. The region has a continental climate that combined the east European climatic character with > 18°c and 19°c in typical summer months of July and in typical winter months -1 to 0.5°c with some exceptional years (Jentsch and Klaeber, 1992). But presently the climatic situation has been changing with some extreme winters of -17°c to -20°c and some very hot summer of 25°c to 40°c, which can be justified from the climatic conditions of the years 2009/2010.

The emission level compared to the average in Brandenburg (MLUR, 2003), fall within the same range with 3.1 kg NO_3/ha a year and 4.2 NH_4 /ha a year, and mean rain water pH of 4.9 (LUA, 2004). There are low industrial activities that cause huge amount of emission in the BR. But the migration effects from coal power plant in Lausitz region and other types of industries that are emitters near the region are sources of emission. Dust that comes mainly from agricultural activities and the pollutant emitted in the atmosphere from different sources like tertiary activities also threaten the air quality in the region. Anthropogenic activities have also influenced the development of the biodiversity of the region (Krausch, 1960). That is almost all the socio-culture and economic activities taking place in the region can be trace from the tradition of the first settlers in the area with some few modifications.

4.2 Data Collection System and Aggregation for Estimating Ecosystem Services

The data sampling strategy and aggregation method for ES that was established in section 3.2 was applied at the UNESCO BRS. The data collection comprises of an observatory survey of the categories, components and indicators of ES under the various types of services, factors influencing their state as was established section 2.1. The influential factors like during the sampling period there were high rains that caused floods in the region, strong wind that bring down trees and negative spread effect from nearby areas. The nearby areas around Straupitz (Caminchen, Sacrow and Waldow) are sandy region blowing sand flakes in the BR. This may cause some of the variables to change their original state at the time of sampling, which may have rendered the observable decision not reflective and realistic. Therefore the opinions of people living around these areas were important for mitigating internal and external influential effects before deciding on the rank value. In a main region chosen for data collection, areas were marked from it as designated units for data collection comprising of smaller units in a county such as city, community or village. These designated areas were chosen using the internet satellite map based on coverability to achieve sample representativeness, which was constantly adjusted for quality management proposes. That is why after every sampling day, the designated sample area for the next sampling day has to be considered to avoid repetitiveness of sampled data. The data were collected in a systematic way by consideration one designed unit at a time and making a survey through it, then filling the working sheets based on value judgment using the ranking scale (see table 9) before considering another unit. The value judgment for each type of ES on a unit area of measurement was derived based on the parameters for indicator system (see Section 2.1). That is observing the categories, components, and natural and human activities that impact them (see table 1 and 2), then taking the decision for a rank value with respect to their availability, pressure and adaptive states (see section 2.1) to support the indicator (see table 3).

The first try was carried out in the Kolkwitz community (the part belonging to the UNESCO BRS) as a pre-test of the sampling strategy, which then processed to the whole of the BR. Data were first collected in the Upper UNESCO BRS and then follow by the Under UNESCO BRS, which provided a huge dataset. The main areas of measurement at the Upper UNESCO BRS were within Lübbenau, Burg and Straupitz, while at the Under UNESCO BRS were Lübben, Schlepzig, Krausnick, and Märkische Heide. The following are designated units of measurement under these main areas:

- Lübbenau: Lübbenau, Südumfluter, Vetschau, Naundorf, Raddusch, Märkische Heide (Vetschau), Leipe, Lehde, Krimnitz and Ragow
- Burg: Müschen, Burg, Burg Kolonie, Burg Kauper, Kolkwitz, Millersdorf, Brahmow, Babow, Kunersdorf, Papitz, Werben and Schmogrow
- Straupitz: Nord Umfluter, Straupitz, New Zauche, Alt Zauche, Byhleguhre, Burglehn, Wußwerk, Radensdorf and Mühlendorf
- Lübben: Lübben, Hartmannsdorf, Groß Lubolz, Klein Lubolz and Bömischen
- Schlepzig: Petkamsberg (3 units) and Schlepzig (2 units)
- Märkische Heide: Märkische Heide, Alt Schadow, New Schadow, New Lübbenau and Kuschkow
- Krausnick: Krausnick, Krausnicker Berge, Groß Wasserburg and Leibsch

Data from several units of measurement were aggregated as a main unit (main units like Lübbenau, Krausnick, Burg) and then to the whole of the UNESCO BRS to suit the analysis.

This was done using the aggregation procedure established in the methodology (see section 3.2.2). For example in the administrative unit of Straupitz to derive the value for the provisioning type of ES, first their category where considered (Food, water, fibre, wood, energy and bio-chemicals). In each category, examples exist for determining them. Then their component and natural human activities that impact them were used to decide on the state of the parameter of the indicator system. Then the rank values were based on the indicator system, categories of ES and activities that have an impact on them (see tables 1, 2 and 3), which can also be verified from landscape identification (see section 4.1). 9 chosen sampling units were randomly observed within areas of differences in km^2 (see table 7) and given rank values. The values were aggregated to give the value for the provisioning service of ES at Straupitz. That is applying the aggregation formula established in section 3.2.2 gives the value 2.0 and it was converted to a percentage to give 40% (see result in figure 19). That is where x_{1-9} stand for the rank value for each of the sampled areas, then it was computed as:

$$\sum_{i=1}^{9}(4x_1 + 9x_2 + \ldots 16x_9) / \sum_{i=1}^{9}(4 + 9 + \ldots 16)$$

The same was done for computing the value of provisioning service of ES for all the other administrative units at the UNESCO BRS and aggregating them together by simply taken their arithmetic mean to derive the sum for the whole. The results are converted and presented as percentages. These same computing steps were also applied for the other ES types. Then summing up the different aggregated types of ES (provisioning, regulating, supporting, preserving and cultural services) to derive the value for an administration unit or Upper and Under UNESCO BRS or the whole of the BR. Data for potential demand and supply for ES were also computed in the same way. This makes it possible the presentation of estimates of ES from smaller units to higher administrative levels.

4.3 Results of Estimates of Ecosystem Services

Land surface, biodiversity and water was used as the main landscape components for estimating ES in the UNESCO BRS that were then aggregated. Therefore using the computation procedure explained in section 4.2, estimates for ES were derived. Figure 19 presents estimates of the different types of ES at the UNESCO BRS with the different lines indicating the different types.

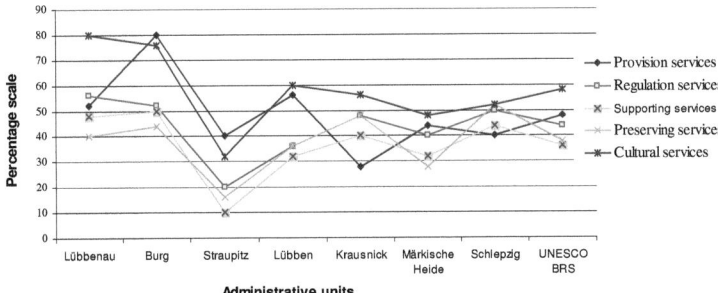

Figure 19: Estimate of Ecosystem Services in UNESCO Biosphere Reserve Spreewald

They can be interpreted using the parameter estimates (see section 3.3). The x-axis shows the different administrative units, while the y-axis is scale to percentages reflecting the values of the converted data (see table 9). The variables of the data can be depicted by matching them to spots corresponding to their percentage value on the y-axis and the administrative units on the x-axis. But the full lines are not trends. Provision services in Burg is 80% (high relevant capacity), while in Straupitz is 40% (low relevant capacity). It shows that the provision services are in high relevant capacity in Burg. But more than very low relevant capacity in Krausnick and for the whole of the UNESCO BRS, they are in more than low relevant capacity. In Burg there are enough water supplies, greenlands and herbs for cosmetics, intensive agricultural activities, fibres, wind power parks. In Krausnick, especially at Krausnicker Berge, it is a hilly area with insufficient water supplies, poor soil and less agricultural or greenland and herb. But the area around Groß Wasserburg and Leibsch, there is water availability more for recreation than agriculture or other activities that leads to provisioning services. While on the other hand, the cultural services are at a high relevant capacity in Lübbenau and almost at a high relevant capacity in Burg. Lübbenau and Burg are the highest attractive places in BR for recreation with high aesthetics, preservation of the local tradition, parks and huge investments in tourist facilities. The cultural services are at a more than very low relevant capacity in Straupitz because of the very low relevant capacity of all the other services. Straupitz has less aesthetics and recreational activities due to the increase degradation of the landscape and less tourist attraction leading to less investment in tourist facilities. Even though, the people preserved their tradition, but it is insignificant due to lack of aesthetics value of the landscape. Therefore there is very little investment in other cultural services and the population that preserved their culture are increasingly immigrating out of this administrative unit. The other administration units and the aggregate of the BR show that the services are within very low and low relevant capacity. They are in high relevant capacity in some units and low in others, but the results show that they are in low relevant capacities in many units in the BR. This means that there is an environmental deficit in the region due to a sink in ES. The demands and supplies of ES for estimating their flow over the BR were considered. They were estimated from the main landscape components based on the natural and human activities that can potentially lead to their decrease or increase at a landscape scale (see table 2). They were considered under potential market services such as carbon sequestration, biodiversity and water resource as MES (see sections 2.3.2, 2.3.3 and 2.3.4). Data were collected on the activities that can potentially lead to their demand and supply at the BR and aggregated as in section 4.2 to derive values for the various administrative units. Figure 20 represents potential demands and supplies for carbon sequestration within the different administrative units of the BR. It shows that the demands and supplies for carbon sequestration for MES have no relevant capacity in the whole of the UNESCO BRS, under UNESCO BRS, Burg, Krausnick, Lübben, Märkische Heide and Schlepzig. In Lübbenau, they have a very low relevant capacity and also a very low capacity in the demand in Straupitz. This also relates to the landscape characteristics, less green areas, forest and biological species for carbon sequestration in Straupitz requiring a higher demand for them, while in Lübbenau there are available.

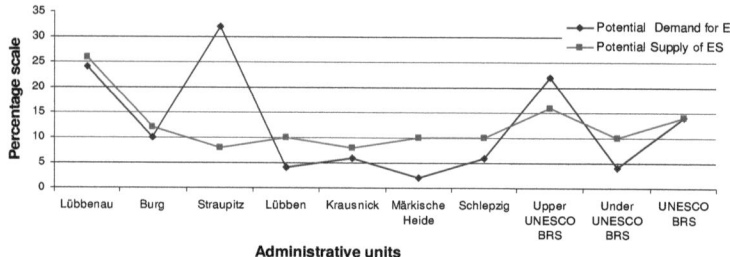

Figure 20: Potential Demand and Supply of Ecosystem Services for Carbon Sequestration

But also a lot of economics activities are going on there that demand them like the tertiary activities, farm machines and the activities of the population. Therefore for policy targets or measures in achieving scenarios for balancing the demand and supply for carbon sequestration in the UNESCO BRS, the stock of ES for carbon sequestration activities have to be ≥ 0. That is their supply must be equal or greater than zero (see section 3.3). But for policy measure or targets on environmental balancing for carbon sequestration, at least the medium relevant capacity have to be achieved. Therefore the supply for carbon sequestration has to be greater than the demand to a level that would have a medium relevant capacity for targets on environmental protection. Data were also collected for the demand and supply for biodiversity. Figure 21 represents the potential demand and supply for biodiversity within the different administrative units of UNESCO BRS. It shows that the supply of biodiversity for MES has no relevant capacity within the administrative units and the whole of the BR, except in Lübbenau with a very low relevant capacity. The UNESCO BRS is a degraded landscape protected to preserve the decreasing biodiversity with less investment for rehabilitation activities. It is reliance on natural dynamic (UNESCO concept of landscape protection) for regeneration of biodiversity, which is very slow with some areas being resilence like Lübbenau. On the other hand the demand shows a medium relevant capacity in Straupitz with a low relevant capacity in Upper BR. While in the whole of the BR, Burg and Lübbenau there is a very low relevant capacity. Areas with poor soil condition and biotic resources would need more biological diverse habitants for genetic resources, bio-refugia, insects and micro-organisms for borrowing activities in soil, bee for pollination, and mitigation of land fragmentation (see table 2). Therefore just like carbon sequestration, for achieving environmental balance the supply of biodiversity has to be greater than the demand to a level that would have a medium relevant capacity.

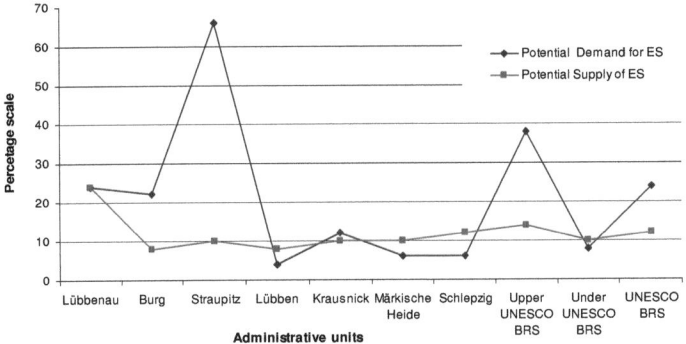

Figure 21: Potential Demand and Supply of Ecosystem Services for Biodiversity

But for the case of Straupitz, the supply has to be high to just balance the demand. Data were also collected for the demand and supply for water resource. Figure 22 represents the potential demand and supply for water resource within the different administrative units of UNESCO BRS.

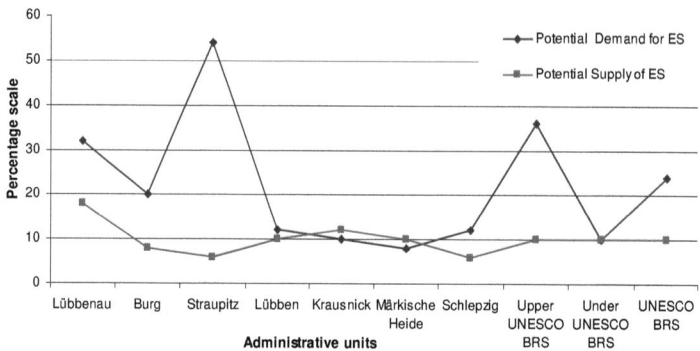

Figure 22: Potential Demand and Supply of Ecosystem Services for Water Resource

It shows that the supply for water resource for MES has no relevant capacity in any of the administrative units and the whole of the BR, except in Lübbenau with almost a very low relevant capacity. Water flow are not managed to provide a huge amount of hydrological services such as control of euthrophication (water quality), control seasonal flow (water quantity), water erosion that were given in section 2.3.3. While the demand shows an almost medium relevant capacity in Straupitz and a very low relevant capacity in the whole of the BR and Lübbenau, but a low relevant capacity in the Upper BR. Straupitz is a dry area with insufficient water supply and when it rains all the water is runoff because of no living soil, especially in areas like Wußwerk, Alt Zauche, Neu Zauche. This can be also noticed within the people in these areas who are resilience to water resources by having tanks for water storage. Therefore, just like carbon sequestration and biodiversity for achieving environmental balance for water resources, the supply have to be greater than the demand to levels that would have a medium relevant capacity. But in the case of Straupitz, the supply has to be high to just balance the demand. Data was also collected on action oriented activities that were assumed to reflect the various actors that were given in the Petri net (see figure 10). There were some activities that were regard as consumption, supplying, developing, producing and antagonising ES, and the action of tourists. For example activities like reducing forest land for intensification of farming was regards as consumption, irrigating land was developing, canalisation was producing, agricultural production was supplying, land fragmentation was antagonising, tourist trumping on green plants was migrants and many more. The data was collected using the same ranking and aggregation procedure, but the decision on them was based on the frequency of activities. Then the values are also converted as percentage. In table 10 the rate of demand and supply of ES by multi-agent is presented. It shows the rate of demand and supply of ES by multi-agent as percentages. These rates of demand and supply for ES also reflect the results of the demand and supply for MES for UNESCO BRS established in this section (see figure 19 to 22). This can be explained from the degradation of the landscape and the reduction in the component that provide ES.

Table 10: Rate of Demand and Supply of Ecosystem Services by Multi-agents

Actions-Oriented	Rate of Demand of ES %	Rate of Supply for ES %
Consumption	26	-
Supplying	-	28
Developing	10	6
Producing	20	12
Antagonising	28	-
Tourist impacts (migrants)	18	-

4.4 Modelling and Simulation Results for Managing Ecosystem Services

The data were encoded in a defined Petri net for modelling and simulating the demand, supply and stock of ES. This is to understand the long term trend in the behaviour of the data for design, control and management of future scenarios for preserving ES. Therefore any administrative unit or upper or under BR or particularly ES or their aggregated can be considered for simulation as they were defined in the method for the need of completeness (see section 3 and 4.2). This would only require a specification of their corresponding colour sets, then also encoding their corresponding data of the administrative units to be studied or the particular type of ES (generic model). But here, the model is tailored to the specific case of the whole of UNESCO BRS as aggregates by defining and encoded the resulting data from the estimates of ES in the Petri net. In table 11 a summary of dataset that were encoded in the places of the defined Petri net (see figure 12) is presented. It shows the initial data for places on the Petri net as aggregates (see figure 19 to 22). They are converted to 1500 as was established (see table 9) and encoded on the various places on the Petri net. Modelled data for two scenario application are also given. For the initial value of the "Demand unit for ES", it was gotten by taking the mean results of 3 MES established as for the UNESCO BRS (see figures 20, 21 and 22). That is 14, 24 and 24 respectively given approximately 20. For the initial value of the supply unit for ES was taken also from the 3 MES. That is 14, 10 and 12 respectively given 12. On the other hand for the stock of ES, it was taken from the mean aggregate of all the ES in the UNESCO established (see figure 19). That is 48 (provisioning services), 44 (regulating ES), 36 (supporting services), 38 (preserving services) and 58 (cultural services) giving 45. The aggregated data for the whole of UNESCO BRS are transformed with respect to the aggregated colours to give the value 1500 (see table 7). The number 1500 encoded at the place stock of ES and the imaginary place "MES" are an assumed value of a very high relevant capacity of ES in the region (approximation level).

Table 11: Summary of Dataset Encoded in the Places of a Defined Petri net

Places in Petri net	Initial Data (Mean value in %)	Aggregated data (1500)	Modelled data scenario 1 by increase in supply (%)	Modelled data scenario 2 by higher increase in supply (%)
MES	-	1500	100	100
Demand unit for ES	20	300	20	20
Supply unit for ES	12	180	20	100
Stock of ES	45	1500	100	100

The value for stock of ES (see figure 19) cannot be taken because that was what was established at a time frame without knowing its initial value. But the value 1500 was taken as a complete space for experimental studies of the changes in this value. The place supply unit for ES is encoded with the value 180 of 1500 corresponding to the aggregate of potential supply of ES, while the place demand unit for ES is encoded with the value 300. The place "MES" is logic to the stock of ES that can be a "Market or other space" because it is imaginary. This means the supply of ES has to come from somewhere and a demand for ES has to go somewhere (nature or market) provoked by an action (Agent) as was established in the methodology (see section 3.1) and defined (see section 3.1.3). The transitions express functions for the rate of firing are also encoded in the Petri net, but can only be visualised in the background when performing a simulation of the model. They contain the rates of occurrence corresponding to multi-agent potential demand and supply of ES. The data on the rate of demand and supply of ES by multi-actors that was established (see table 10) is encoded in the various transitions respectively. Then the weight on the edges (arrows) were maintained as defined and executed (see section 3.1.4) for the through flows of tokens (capacity of colour set), when the guard evaluates (TRUE – flow or FALSE – no flow) in accordance with the defined kinetic rule (see section 3.1.3). The simulation results are generated based on computation steps of eliminations of linear combinations on the incidence matrix that was defined in section 3.1.3 and their probabilities accordance to Gillespie (stochastic simulator in Snoopy software) defined in section 3.1.4 on the rate of transition. This is to generate a one simulation count in which a stepwise 100 runs or more can be generated (specified). For solvable systems a matrix is give as "Ax = b", where here "A" is the incident matrix (A_i), "x" is the vector (p_1, p_2, p_3, p_4) and "b" the resulting vectors lines (b_1,

b_2, b_3, b_4) on a plane, which through elimination the particular solution is obtained (Gaussian elimination). If it is assumed that the stochastic probability (Gillespie) function is set on the transition rates T ($t_1, t_2, ..., t_8$) is ϖ:

$$\varpi A_i x = b \tag{15a}$$

$$\varpi \begin{pmatrix} 0 & 0 & -1 & 1 \\ 0 & 0 & -1 & 1 \\ 1 & 0 & -1 & 0 \\ 1 & 0 & -1 & 0 \\ -1 & 0 & 0 & 1 \\ -1 & 0 & 0 & 1 \\ 0 & -1 & 1 & 0 \\ 0 & 1 & 0 & -1 \end{pmatrix} (p1\ p2\ p3\ p4) = (b1\ b2\ b3\ b4) \tag{15b}$$

$$\varpi \begin{pmatrix} 0 & 0 & -1 & 1 \\ 0 & 0 & -1 & 1 \\ 1 & 0 & -1 & 0 \\ 1 & 0 & -1 & 0 \\ -1 & 0 & 0 & 1 \\ -1 & 0 & 0 & 1 \\ 0 & -1 & 1 & 0 \\ 0 & 1 & 0 & -1 \end{pmatrix} (1500\ 300\ 180\ 1500) = (b1\ b2\ b3\ b4) \tag{15c}$$

The value for the vector "x" is the aggregated data of the initial values (see table 10) representing for number of tokens in the corresponding places in the Petri net. Then assuming ϖ (Snoopy in build) to the transition rate (see table 10) and using all the defined initial stage to run the simulation of 100 counts to obtain "b". "b" is the resulting simulation solutions at the place MES (b_1), Demand unit for ES (b_2), Supply unit for ES (b_3) and Stock of ES (b_4) using the Petri net (see figure 12). In table 12 a summary of simulation results with the data of ES from UNESCO BRS is presented. Only b_2, b_3 and b_4 are given here since the interest is only them.

Table 12: Summary of Simulated Results Generated with Data of ES from UNESCO Biosphere Reserve Spreewald

Simulation Counts	Demand unit for ES	Supply unit for ES	Stock for ES
0	300	180	1500
1	422	104	1386
2	650	66	1059
3	832	85	747
4	1043	51	492
5	1106	75	315
.	.	.	.
.	.	.	.
10	1018	64	53
15	581	54	48
.	.	.	.
.	.	.	.
100	76	33	48

It gives a summary of the results that were generated. That is the initial stage (0), then the first 5 simulation counts, then 10, 15 and the 100 count for the demand unit for ES, the supply unit for ES and stock of ES. But the results are plotted at an interval of 5 simulation count (see figure 23) for presenting the trend of ES at the UNESCO BRS. Figure 23 represents the simulation result of the net to study the trend of ES at the UNESCO BRS. The mean aggregated values for ES are on the y-axis, while the number of simulation counts is on the x-axis that is assumed as the time for observing ES at the UNESCO BRS over a long time.

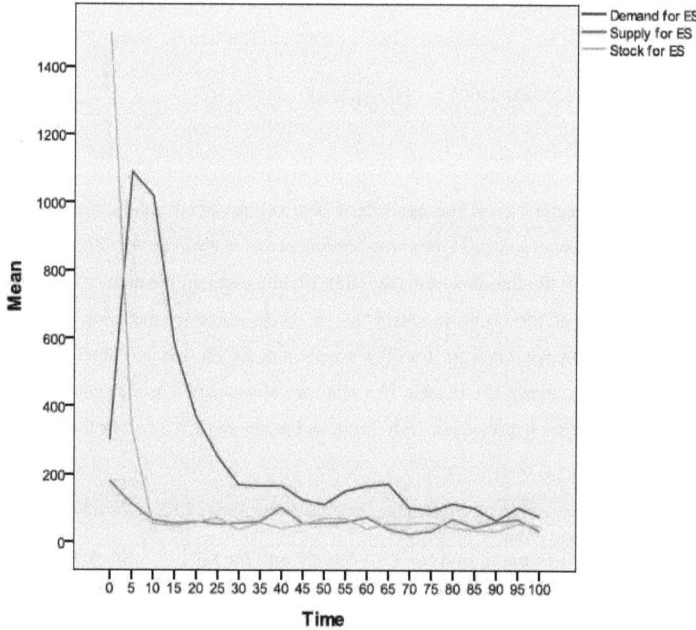

Figure 23: Trend of Ecosystem Services at the UNESCO Biosphere Reserve Spreewald

It shows that the very high relevant capacity of stock of ES in the region that was assumed in the data would drop sharply because demand for ES is drastically increasing, while their supply is at a no relevant capacity (deficit balance). Then as the demand for ES increased shortly to almost a high relevant capacity, it would start to drop quickly because there is no relevant recharge from their supply that remains at no relevant capacity throughout the trend. This would render the high stock of ES to quickly drop to a no relevant capacity and remains at that level throughout the trend. The decreasing demand for ES would also reach a no relevant capacity. Therefore all of them would remain at a no relevant capacity throughout the rest of the trend. This means a typical scenario trend for high degree of landscape degradation that would be very vulnerable to the people in the region. This can be an early warning for the need of measures for mitigation, rehabilitation and restoration of ES at the UNESCO BRS. Therefore alternative scenarios are modelled for studying the different variables that can be of target for measures of potentially achieving environmental improvement and sustainability.

The data of the input variables in the Petri net that were established (see table 11) are manipulated (data mining) for empirical determination of scenarios for environmental improvement. Figure 24 presents a modelled trend for management of ES by slightly increasing the supply of ES from a level of no relevant capacity to very low relevant capacity (use: modelled data scenario 1 by increase in supply in table 11). The y-axis and x-axis mean the same as was explained (in figure 23). It shows that the trend in the stock of ES would remain almost the same as the formal result.

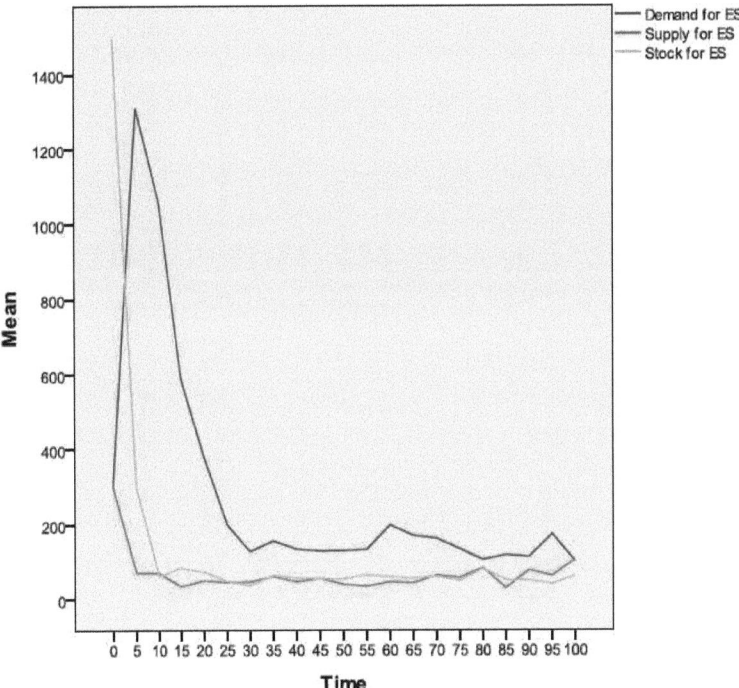

Figure 24: Modelled Trend of Ecosystem Services by Slightly Increased in Supply

While the demand for ES would increase slightly above the high relevant capacity, but more than the formal result (see figure 23) due to the increased in their supply and would drop again very fast. This is because the capacities of their supply are irrelevant to meet the demand for ES, which then have the same behaviour in the trend as the formal result. A further study was done on the supply of ES by increasing it to the level that just equates the high assumed stock of ES (use: modelled data scenario 2 by higher increase in supply in table 10). Figure 25 presents a modelled trend by increasing supply of ES from a level of no relevant capacity to very high relevant capacity. The y-axis and x-axis also mean the same as was explained (in figure 23). It shows that the very high capacity of the supply of ES would immediately fall back to a no relevant capacity, while the stock of ES would slightly increase shortly. But it would drop back drastically also to a no relevant capacity. The demand for ES would increase highly above the very high relevant capacity because of their increased in supply due to the high need for them.

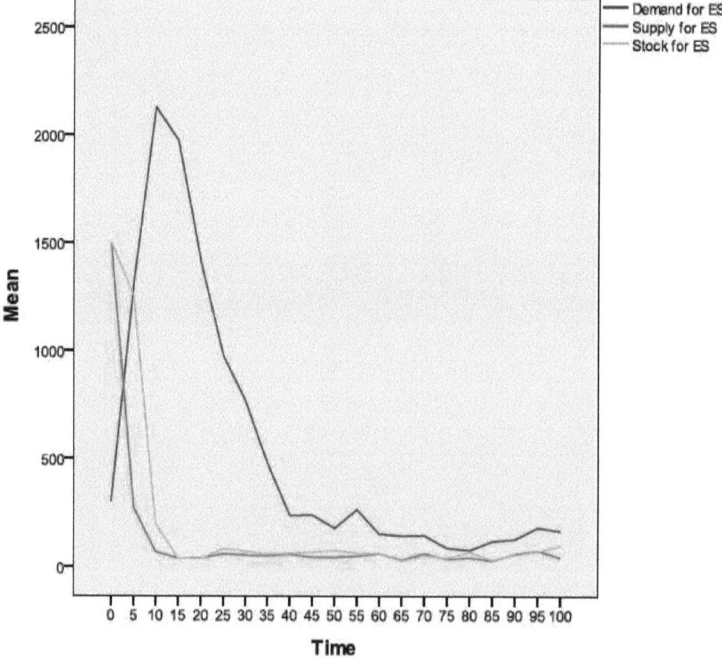

Figure 25: Modelled Trend of Ecosystem Services by Higher Supply Level

Then they would drop back again drastically to a no relevant capacity and remain almost constant at that level throughout the rest of the trend due to no reaction from their supply to recharge them. This means environmental improvement measures by increasing the supply for ES would have no major effect. But it makes no sense to model the demand for ES because their relationship depends on one another, so the demand and supply of ES by multi-actors activities were modelled by manipulation of their values in the transitions. This was to study from the model whether controlling the multi-agent can have any impact on the results. Then the values of the rate of demand and supply of ES by multi-agents that was established (in table 10) were all increased to 100. Figure 26 presents a modelled trend of ES by slightly increasing the value of multi-actors actions for the supply/demand of ES from a level of no relevant capacity to very low relevant capacity. The y-axis and x-axis also mean the same as was explained (in figure 23). It shows a great change in the stock of ES and the supply of ES in which all remain within a lightly very high relevant capacity and more than a very high relevant capacity.

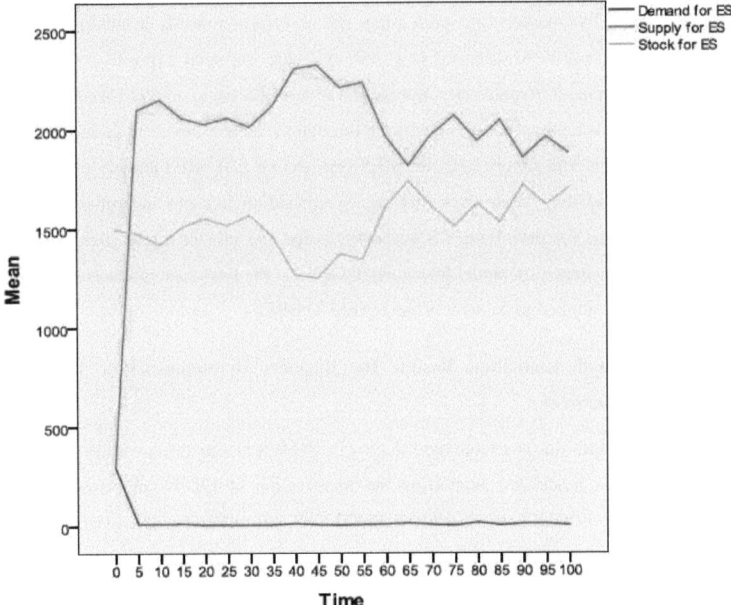

Figure 26: Trend of Ecosystem Services by Studying Multi-actors Actions

This means when changing the value based on multi-actors actions on their supply or demand rate for ES in the Petri net, immediately their supply increased very sharply to a more than very high relevant capacity. That is the supply of ES would increase from a very low relevant capacity to a far more than very high relevant capacity. While the stock of ES would increase in respond to the very high supply of ES and remain at a more than very high relevant capacity throughout the rest of the trend. But supply of ES would drop again to a medium relevant capacity and remains between a more than medium and high relevant capacities.

On the other, the demand for ES would immediately drop from a very low relevant capacity to a no relevant capacity and remain constant at that level throughout the rest of the trend. This can be interpreted as; when there is environmental improvement, multi-actors would demand only very less than no relevant capacity as a result of insignificant landscape problems. This means soil, water and land surface are in good state due to good preservation measures that are being put in place and multi-actors are responding positive to them. Therefore there is the need for a simulation model that can provide a cooperation strategy with multi-actors for markets for preserving ES as efforts towards achieving such an environmental improvement scenario. One way of achieving such a positive scenario is through the development, maintenance and preservation of ES using market-based strategies. That is business development for preserving ES through the establishment of trading platform for potential demand and supply to foster MES (see section 2.3). MES already existed with other emerging and innovative ones that the Spreewald region can incorporate in their management scheme for preserving ES at the landscape and protect nature (Fongwa et al., 2011). Therefore a simulation model for managing MES at the landscape that was established (in section 3.1.4) is applied in the case of the UNESCO BRS.

4.5 Modelling and Simulation Results for Business Development for Preserving Ecosystem Services

A management framework to encourage the growth of MES is structured with Petri net as a business simulation model for supporting the preservation of ES. It comprises of the 3 modelling frameworks with Petri nets (see figures 12, 13, and 14) and coupled (see figure 15) as a unified system for managing ES at a landscape that was established in section 3.1.4. It is designed as a decision support system for options on MES for potentially achieving scenarios of environmental improvement and sustainability of landscapes. In the unified net (see figure 15), the bottom net contains markings of the left net. That is, it has colour sets for MS and PP, and the flow is defined as n ++ and (x, y). The flow is a combination of the two colour sets (token of ES). On the other hand, the net on the left has marking of financial flow. Therefore the place where the nets are coupled has all the various types of markings that make up a unified net. The nets are specified the same as that for the improved scenarios for landscape management of ES (see figure 26) and the CFP and management to support to achievement of balancing ES at landscape and growth of MES at the UNESCO BRS. Therefore for application of this unified Petri net modelling system in the UNESCO BRS, the data systems also need to be specified and put in control variables for achieving management to support the growth of MES and balancing ES at the landscape.

The data of the defined Petri net for modelling multi-agent activities for estimating ES (see figure 12) that is explained in section 4.4 is maintained the same here. This is the right side of the unified Petri net (see figure 15). The data for the defined Petri net for modelling management scheme for MES (see figure 13) and the defined Petri net for modelling CFP for developing MES, maintained the initial theoretical dataset as was established (see figures 13, 14 and 15). This is based on their colour set as tokens, which are basic requirement of their input data, since there were no data for them from the UNESCO BRS. The defined Petri net for modelling management scheme has a 2x3 matrix and that of CFP has a 7x3 matrix. Their incident matrices are established as that of the defined Petri net for modelling multi-agent activities for estimating ES (see figure 12) given in section 3.1.3 and also computable as in section 4.4. They can be computed also with their incident matrixes. But the tokens in the places (see figures 13, 14 and 15) are given a higher value by multiplying them by 100. All their rate of transitions are also given 100 each as was seen as a good approximation for achieving improved scenarios (see figure 26) explained in section 4.4. This was to match the values of the variables to correspond to the data system established in section 4.4 (see table 11), since they need to be coupled with data that has another scale. Therefore it was necessary to bring them to a common scale such that it can ease their interpretability. The data set were arbitrary improved for the left side Petri net that has the CFP scheme in the unified net, and also a control restriction was put at the arc that lead from the place "Market" to " Flow of benefits" (see figure 15). The weight was doubled (2`a) in order to control that all flow of contributions to financing should not become benefits. This could lead to a constant stream of benefits to financial contributors and a constant financing (finance) at their disposal to either save or still contribute to financial portfolio. Then their places were arbitrary adjusted with different values. The place "Flow of benefits" was multiple by 6 given 2 x 100 x 6 = 1200, "Financing" by 3 given 2 x 100 x 3 = 600 and "Contribution unit" by 5 given 2 x 100 x 5 = 1000 as their own initial values to start the computation. In table 13 a summary of the unified simulation results of the computation for management support system of MES at the UNESCO BRS is presented. It shows the initial values of the variables under studies (Simulation count of zero) before and after the simulation runs to 100 counts.

Table 13: Summary of Unified Simulated Results Generated for Management support System of MES at the UNESCO Biosphere Reserve Spreewald

Simulation Counts	Demand unit for ES	Supply unit for ES	Stock of ES	Financing	Benefits	Contributions to financing	Market for management scheme	Market for ES
0	300	300	1500	600	600	1000	1800	2000
1	15	2134	1443	600	600	2104	1800	504
2	4	2134	1422	600	600	2008	1800	604
3	15	2105	1473	600	600	1906	1800	701
4	8	2133	1451	600	600	1810	1800	798
5	11	2087	1498	600	600	1700	1800	904
.	.	.	.	.	.	.	.	.
10	13	2109	1472	600	600	1206	1800	1400
15	9	1759	1823	600	600	634	1800	1973
.	.	.	.	.	.	.	.	.
100	12	2126	1454	600	600	0	1800	2608

The first 5 counts are given and then the data were collected after 5 complete simulation counts till up to 100 counts. Figure 27 represents the simulated results for managing markets for preserving ES at the UNESCO BRS that were generated in an interval of 5 counts. The y-axis and x-axis also mean the same as was explained (in figure 23) and the various curves of the variables under studies (see table 13) as their trends. The simulated result for the demand unit for ES, supply unit for ES and stock of ES remain the same as before (see figure 28) as can be seen in the results. On the other hand, the results show that the more than very high capacity of financial contributions from stakeholder would drop because the funds would have been invested in the maintenance and development of ES. The high relevant capacity of benefits that aroused from MES (NPV projects) due to the high demand for ES assumed, would then drop to a low relevant capacity and remains almost constant. But financing that was used to contribute for MES would be covered up immediately (short pay back time) and then remain constant at a low relevant capacity as before.

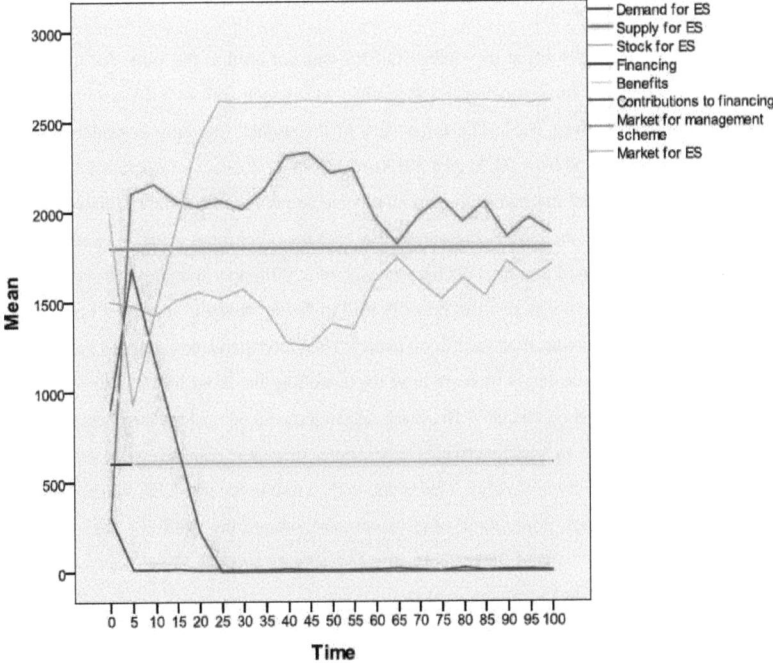

Figure 27: Simulation Results for Managing Market for Preserving Ecosystem Services at the UNESCO Biosphere Reserve Spreewald

Transaction for market scheme would remain constant at a more than very high relevant capacity, while the markets that have developed would grow to a more than very high relevant capacity. Then it would drop again to a little below the low relevant capacity, due to lack of experience. Then when experience is being developed within these markets, they would grow to a more than very high relevant capacity and remain constant. This scenario for market scheme and MES can lead to an integration of the community leading to job creation and poverty reduction in the Spreewald region when applicable (improvement in social welfare). It can also encourage the main inhabitant in the region to remain in the region because they would have employment (taking jobs to the population as economic strategy to management demographic problems). Therefore regional product, personal management, energy, industrial parks and other sustainability ideas that are within the discussion of the Spreewald Union (Spreewald Verein) can then hold effective grounds.

4.6 Discussions of Results

The value estimated for ES at the UNESCO BRS was not used as the value for the stock of ES in the modelling for estimating ES (see table 11), which was 45% (when converted to 1500 would have given 675). The input data to the model was only considered as at a particular event (period from 02 August 2010 to 30 October 2010). The stock of ES cannot be assumed as zero before that transition period, it could be plus or minus the estimated value for ES. So for estimating the long term trend for ES with the other input values (see table 11 and 12), the stock of ES was assumed the higher value of 1500 (upper triangle). The demand and supply of potential activities that can possibly lead to the demand and supply of ES were also estimated at that particular time (transition event). Therefore the values are only valid for that particular observation event, which are used for modelling the trend of ES for a longer time for understanding of their changes. There are also trajectories in ecological processes for ES, which are interwoven in which activities easily impacting their demanding or supplying ES (rates) may not be easily noticeable. This is due to the complexity involved, which could only be understood through long term observation and monitoring studies. They were not considered only the demand and supply oriented actions that could be traced out at the landscape of the UNESCO BRS were taken.

The number of multi-actions (multi-agents) oriented activities that demand ES were the same as whose that supply them, but their rates were not the same (see table 10). The results (see figures 23, 24, 25 and 26) of the modelling and simulation using the established values for ES (see figures 20, 21 and 22,) and their rate (see table 10 and the assumed value of 100 for all) can provide an important understanding that strategies to preserve ES should not only rely on increasing the supply of ES. This can be justified with the simulation results of slightly increasing the value of the supply of ES to a higher level (see figure 24) and again to the maximum level highest level (see figure 25). It provided no major change in the simulation results as before (see figure 23). But when increasing the rates of demand and supply of ES by action oriented activities (see figure 26), there was an increasingly higher capacity for the supply and stock of ES. Even thought the demand for ES falls or have no significant change. This is logical due to the fact that when the supply of ES would increase, it will also lead to the increase in the stock of ES. But the demand for ES needs to fall due to insignificant activities on the landscape that requires ES to balance them for environmental improved scenarios. Therefore management strategies for balancing ES at the landscape should implement control measures to impact on multi-action oriented activities by multi-actors. MES and CFP are introduced as a management option that can potentially be use as a management strategy for balancing ES. The assumption is that a model of market scheme for balancing and preserving ES at the landscape would encourage an increasing rate of supply of ES and reduction of the rate of demand for ES. A model of CFP would encourage the participation of multi-stakeholders in the MES, which would encourage these markets to grow. Therefore for a management support system; they are coupled with the estimating model for ES. The model for market scheme for ES and CFP were only experimentally demonstrated in theory within the model for estimating ES. Therefore it can be considered as just a proposal for options to manage ES at the landscape. It still needs to be tested and verify with real scenario data that were not available at the UNESCO BRS.

5. Conclusions

Ecosystems are complex systems with a number of components with different processes and functions providing ES. But even though they are complex, determining their process and functional units that are linked to the sink or improvement of the supply of ES on the landscape can provide an understanding of their dynamics. Then it would be possible to estimate them for setting measures that would potentially lead to their preservation. Therefore data sampling and management are essential for analysis and structuring strategic measures for preserving ES. This has been realised in this thesis within the data sampling on ES, modelling and simulation frameworks presented. The challenges in nature protection and preservation of ES are the identification of appropriate units for understanding and implementing measures that can potentially lead to their sustainability. This requires that studies on landscape management for balancing ES at the landscape should defined appropriate units of observations and then aggregating measures for strategic or policy measures. This was presented in the data sampling procedure in this thesis.

The theoretical background, methodology and the results of the modelling and simulation framework at the BRS were able to answer the research questions that were declared in the introduction. Literatures were reviewed to provide a synthesis of indicators for ES (see section 2.1) that a method was established from them for estimating ES. A modelling framework using Petri net (see section 3) for management of ES at the landscape scale was presented. This was applied at the UNESCO BRS to show how it can encourage business development for preserving ES (see section 4). Business development for preserving ES was discussed in the theoretical background (see section 2.3). It was then incorporated in the Petri net modelling framework as market-based strategy for balancing ES (see section 3) and applied at the UNESCO BRS (see section 4). CFP was discussed in the theoretical background as incentive and motivation scheme to encourage multi-stakeholder involvement in business development for preserving ES (see section 2.5) and integrated in the modelling framework in the methodology (see section 4). It was also applied in theory in the UNESCO BRS (see section 4), due to lack of appropriate data for its real scenario analysis that need further verification in future research. The modelling framework was established in the methodology as a unified system for estimating ES on the one hand. On the other hand, providing management scheme for balancing ES at the landscape that couple a CFP framework as incentive and motivation for cost-benefit sharing based pareto optimality. The assumption is that its implementation would encourage increasingly number of multi-stakeholders involvement in business development for preserving ES (see section 3). This was also applied at the UNESCO BRS in theory as a management support system for managing ES at that landscape. Even though quality assurance criteria were point in place, but there were some limiting factors that need to be mention as hints that can be helpful for future research in this field:

- There was the need on brain steaming on landscape components for ES and indicators with other scientists or students on field site to reduce subjectivity in the results. Especially if they have other interdisciplinary knowledge. Even though many workshop and conferences were attended on brain steaming on them, but practical discussion of them on field site was lacking that can be a hint for future research.

- Ecosystems that provide ES need a long term observation and many measurements because they are natural processes with slow dynamic rates. This was not possible within this research because it was a short term research for 3 years that need the development of concept, methods and then data collection. Therefore long term researches are required for studying ES.

- Optimisation techniques can also be essential for complexity reduction in the Petri net modelling framework that was not considered. It was not within the research scope. But if taken into consideration it could be able to achieve a potentially optimal level between complexity and abstraction, which the way forward could be synchronisation using process algebra. This can be considered in future research for improving the quality of modelling of complex systems with Petri net, especially ES.

- Decision variables for management options can also be considered for optimisation of the modelled values for decision support systems, which was not also within the scope of this research. Therefore for future research a management science approach with linear algebra can also be away forward for integration decision variables to achieve optimal solutions.

- The research results need to be compared with other observation and estimates of ES from other region in the world. Even thought the case of agricultural management in Wielkopolski region in Poland (another protected area in Europe) was considered within the research framework, other landscape types and in other continents were not included. Therefore for future research this aspect needs to be considered.

- A survey on CFP and the management scheme were not considered. They need to be considered in future research for real scenario application, especially for practical implementation.

The results of this thesis, particular the case of UNESCO BRS can be used to argue that the establishment of protected areas by the UNESCO administration should not be the only solution to fight land degradation. This is because the regeneration capacities of areas that have a sink in ES are weak. For instance if exposed to natural condition like changing weather pattern (high rain or dryness), then the levels of degradation would increase and also cause threats to the people. Therefore mitigation strategies like restoration or rehabilitation of degraded landscape should be the first steps. Then the implementation of protection measures to encourage the sustainability of ecosystems. This thesis analysed the situation by establishments of service units for studying changing in ES based on their demand and supply through the influence of multi-agents. It provides arguments for market-based strategies for preserving ES on the landscape and beyond. That is by providing a review of different MES and their possible trading platforms, and established a management scheme comprising of certification, permit and conservation banking/credit. Therefore a business modelling framework with multi-agent procedure for a CFP was developed and its simulation results shows that market-based strategy for preserving ES is an option and can be achievable.

References

Adam N. R. and Wortmann J. C., 1989. Security Control Methods for Statistical Databases: A comparative study. ACM Computing surveys 21 (4): pp. 515- 556.

Adèr H. J., Mellenberg G. J. and Hand D. J., 2008. Advising on Research Methods. A consultant's Companion. Huizen, The Netherlands, Johannes van-Kessel publishing.

Ajmone M. A., Balbo G., Conte G., Donatelli S. and Franceschinis G., 1995. Modelling with Generalized Stochastic Petri Net. John Willey & sons.

Allan D. J., 1993. Stream Ecology, Structure and Function of Running Water. Chapmann and Hall, London, UK.

Allan J. K., 2005. Marketing Communication: New approaches, technologies, and styles. Oxford University Press.

Alper T. M., 1987. A Classification of All Order-preserving Homeomorphism Groups of the Reals that Satisfy Uniqueness. Journal of Mathematical Psychology 31: pp. 135- 154.

Artuso A., 2002. Biosprospecting, Benefit Sharing and Biotechnological Capacity Building. World Development 30 (8): pp. 1355- 1368.

Backus J. W., 1959. "The syntax and Semantics of the Proposed International Algebraic Language of the Zurich ACM- GAMM. Proceeding of the international conference on information Processing. UNESCO, pp. 125- 132.

Banks J., Carson J. S., II Barry L. N., 1999. Discrete- Event System Simulation. Second edition, Prentice-Hall, Inc.

Banzhaf S. H., 2005. Green Price Indices. Journal of Environmental Economics and Management 49 (2): pp. 262- 280.

Banzhaf S. H. and Boyd J., 2005. The Architecture and Measurement of an Ecosystem Services Index. Resource for the Future, Washington (D.C.).

Barr N., 2004. Economics of the Welfare State. 4th edition, Oxford University Press.

Batini C. and Scannpieco M., 2006. Data Quality: Concepts, methodologies and techniques. Springer Verlag, Berlin-Heidelberg.

Bauer S. and Stringer L. C., 2009. The Role of Science in the Global Governance of Desertification. Journal of Environment and Development 18 (3): pp. 248- 267.

Begg D. and Ward D., 2004. Economics for Business. The McGraw- Hill Education.

Benneman J., 2003. Technology Roadmap: Biofixation of CO_2 with micro-algae. Final Report (70/00092) to the US Department of Energy Technology Laboratory, Morgantown- Pittsburg.

Bisgaard S., 2008. "Must a Process be in Statistical Control before Conducting Designed Experiment?" Quality Engineering, ASQ 20 (2): pp. 143- 176.

Boyd J. and Banzhaf S., 2007. "What are Ecosystem Services? The need for standardized environmental accounting units." Ecological Economics 63 (2-3): pp. 616- 626.

Bradford S. G., Quint N., Shimon C. A. and Michael A. (eds.), 2007. Emerging Markets for Ecosystem Services: A case study of the Panama Canal Watershed. The Haworth Press, Inc.

Bronner G., Oppermann R. and Rösler S., 1997. Umweltleistungen als Grundlage der Landwirtschaftlichen Förderung: Vorschläge zur Fortentwicklung des MEKA-

Programms in Baden-Württemberg. Naturschutz und Landschaftsplanung 29 (12): pp. 357- 365.

Bruijnzeel L. A., 2004. Hydrology Function of Tropical Forests: Not seeing the soil for the Trees? Agriculture, Ecosystems and Environment 104: pp. 185- 228.

Calow P. and Peets G. C., 1994. The Rivers Handbook. Volume 2, Blackwell, Oxford, UK.

Campbell D. E., 1995. Incentive: Motivation and the economic of informatics. Cambridge University Press.

Carson R. T., 1991. "Constructed Markets." In Braden J. and Kolstad, C. (Eds.), Measuring the Demand for Environmental Quality. Elsevier Press, Amsterdam, Netherlands.

Castanet R., Guitton P. and Ratiq O., 1985. An Automatic System for Studying of Protocols: A presentation and critique-based on a work example. PS TV 4: pp. 111- 125.

Castree N., 2003. Bioprospecting: From theory to practice. Trans Inst Br Geogr NS 28: pp. 35- 55.

Challen R., 2000. Institution, Transaction Costs and Environmental Policy. New Horizons in Environmental Economics, Edward Elgar, UK.

Chambers N. and Lewis K., 2001. Ecological Footprint Analysis: Towards a sustainable indicator for business. ACCA Research paper No. 65, Pub. Certified Accounting Educational Trust, London, UK: pp. 73.

Chamberlain D., Fuller R., Bunce R., Duckworth J. and Shrubb M., 2000. Change in the Abundance of Farmland Birds in Relation to the Timing of Agricultural Intensification in England and Wales. Journal of Applied Ecology 37: pp. 771- 788.

Chapin F. S. III, Matson P. A. and Mooney H. A., 2002. Principles of Terrestrial Ecosystem Ecology. Springer- Verlag, New York Inc.

Chiola G., Donatelli S. and Solda G., 1989. Construction and Validation of a Petri Net Model of a Layered Protocol Architecture. In Fourth IEEE Region, 10th International Conference 2004, TENCON 89: pp. 226- 233.

Christensee B. J. and Kiefer N. M., 2009. Economic Modelling and Inference. Princeton University Press.

Costanza R. (ed.), 1991. Ecological Economics: The science and management of sustainability. Columbia Univ. Press, New York.

Costanza R., d'Arge R., de Groot R., Farber S., Grasso M., Hannon M. B., Limburg R., Naeem S., O'Neill R. V., Paruelo J., Raskin R. G., Sutton P. and van den Belt M., 1997. The Value of the World's Ecosystem Services and Natural Capital. Nature 387: 253–259.

Daily G., (Ed.), 1997. Nature's Services: Societal dependence on natural ecosystems. Island Press. Washington D.C.

Daily G. C., 2006. Conservation Planning for Ecosystem Services. PloS Biology 4: pp. 2138-2152.

Daily G. C. and Ellison K., 2002. The New Economy of Nature and the Marketplace: The quest to make conservation profitable. Island press, Washington (D.C.).

Daly H. and Farley J., 2004. Ecological Economics Principles and Applications. Island Press, Washington (D.C.).

Dasgupta P., 1996. "The Economics of the Environment". Proceeding of the British Academy 90: pp.165- 221.

David R. and Alla H., 2005. Discrete, Continuous and Hybrid Petri Nets. Springer, Berlin.

De Barry P. A., 2004. Watersheds: Assessment and management. John Wiley and Sons.

Desel J., 1992. A Proof of the Rank Theorem for Extended Free Choice Nets. In Jensen K. (Ed.), Application and Theory of Petri Nets. Springer Berlin, Lecture Notes in Computer Science 616: pp. 134- 183.

Desel J. and Juhás G., 2001. What is a Petri Net? Advances in Petri Nets, Springer Berlin, LNCS 2128: pp. 1- 25.

EASAC, 2005. A User's Guide to Biodiversity Indicators. European Academies of Science Advisory Council (EASAC).

EC, 2004. Aid Delivery Methods: Supporting effective implementation of EC External Assistance. Project cycle management guidelines, volume 1, European Commission (EC).

EEA, 2003. An Inventory of Biodiversity Indicators in Europe. Technical Report no. 92, European Environmental Agency (EEA), Denmark.

EEA, 2007. Proposal for a First Set of Indicators to Monitor Progress in Europe: European Environmental Agency halting the loss of biodiversity by 2010. Technical Report no. 11, European Environmental Agency (EEA), Denmark.

Eelko H., 2007. Applied Statistics with SPSS. SAGE publisher Ltd.

EFES, 2001. Opinion of EFES on the Commission Staff Working on Financial Participation of Employees in the European Union. Document SEC (2001) 1308 on the 26 July 2001, European Federation of Employee Share Ownership (EFES), Brussels.

Eike B., Raymond D. and Maciej K., 2001. Petri Net Algebra. Springer- Verlag., Berlin.

FAO, 2005. Global Forest Resources Assessment 2005. Published by Food and Agricultural Organization (FAO), Rome.

FAO, 2007. Adaptation to Climate Change in Agriculture, Forestry and Fisheries: Perspective, framework and priorities. Published by Food and Agricultural Organization of the United Nations (FAO), Rome.

Farber S. C., Costanza R. and Wilson M. A., 2002. Economic and Ecological Concepts for Valuing Ecosystem Services. Ecological Economics 41: pp. 375- 392.

Feldman A., 1980. Welfare Economics and Social Choice Theory. Martinus Nijhoff, Boston.

Fongwa E. and Gnauck A., 2009a. Value Creation from Ecosystem Services for Business Development. In Gnauck A. (Ed.), Modeling and Simulation of Ecosystems, Kölpinsee Workshop 2008, Shaker Verlag, pp. 43- 56.

Fongwa E. and Gnauck A., 2009b. Community-based Financial Participation for Developing Ecosystem Services at the Landscape Scale. In Breuste J., Kozová M. and Finka M. (Eds.), European Landscape in Transformation: Challenges for landscape ecology and management, 70 years of landscape ecology in Europe, European IALE Conference 2009, published by University of Salzburg, Austria: pp. 413- 416.

Fongwa E. and Gnauck A., 2009c. Conceptualization of Community-based Financial Participation for Business Development. Journal of Business and Economics (JBE) 1 (2): pp. 204- 221.

Fongwa E., Oana N., Mircea D., Gnauck A. and Wagner G., 2009. Agent-based Discrete Event Simulation of a Community-based Financial Portfolio for Business Development. In Gnauck A. and Luther B. (Eds.), 20th Symposium Simulation Techniques, ASIM 2009, published by Shaker Verlag, Extend Abstracts: pp. 144 – 147 (Full paper in CD: pp. 294 – 303).

Fongwa E. and Gnauck A., 2010. Land Use Degradation in the African Sahel Region: The importance of developing ecosystem services to fight desertification. In Gnauck A. (Ed.), Modeling and Simulation of Ecosystems, Kölpinsee Workshop 2008, Shaker Verlag: pp. 102- 119.

Fongwa E., Petschick M., Gnauck A. and Müller F., 2010a. Measuring the Rate of Changes in Ecosystem Services at the Landscape Scale: Towards decision support systems for improving the ecological performance. In Andrzej Ma. and Andrzej Mi. (Eds.), Implementing of Landscape Ecological Knowledge in Practice, 1st IALE- Europe Thematic Symposium, published by WYDAWNICTWO NAUKOWE UAM: pp. 223- 228.

Fongwa E., Gnauck A. and Müller F., 2010b. A Hybrid Model for Sustainable Development of Ecosystem Services. In Gnauck A. (Ed.), Modeling and Simulation of Ecosystems, Kölpinsee Workshop 2009, Shaker Verlag, pp. 204- 223.

Fongwa E., Gnauck A., Müller F. and Petschick M., 2011. Business Development for Preserving Ecosystem Services. International Journal of Ecological Economics and Statistics (IJEES) 21 (P11): pp. 19- 32.

Fox J., and Nino-Murcia A., 2005. Status of Species Conservation Banking in United States. Conservation Biology 19 (4): pp. 996- 1007.

Freeman A. M., 1993. The Measurement of Environmental and Resource Values: Theory and methods. Resources for the Future, Washington (D.C).

Freemann A. M., 1995. Accumulation and Relative Surplus Value. The IWGVT Mini- Conference 1995, Published by IWGVT.

Friends of the Earth International, 2005. Nature for Sale: The impacts of privatizing water and biodiversity. Published by Friends of the Earth International, Amsterdam, The Netherlands.

George A. M., Louiqa R. and Maria- Esther V., 2000. Using Quality of Data: Meta data for source selection and ranking. In proceedings of the ACMS/GMOD Workshop on Web and Databases (WebDB): pp. 93- 98.

Gilbert J., Henske P. and Singh A., 2003. Rebuilding Big Pharma's Business Model. In Vivo, The Business and Medicine Report 21 (10): pp. 73- 83.

Gillespie D. T., 1977. Exact Stochastic Simulation of Coupled Chemical Reactions. Journal of Physical Chemistry 81 (25): pp. 2340- 2361.

Gillespie D. T., 1992. Markov Processes: An introduction for physical scientists. Academic Press, Boston.

Girault C. and Rüdiger V., 2003. Petri Nets for Systems Engineering: A guide to modelling, verification, and application. Springer- Verlag, Berlin.

Gnauck A., 1988. Kybernetische Beschreibung Limnischer Ökosysteme. Habilitationsschrift Fakultät Bau-, Wasser- und Forstwesen, Techn. Univers. Dresden.

Greenspan- Bell R. and Russell C., 2002. Environmental Policy for Developing Countries. Issues in Science and Technology (Spring): pp. 63– 70.

Grime J. P., 2007. Biodiversity and Ecosystem Function: The debate deepens. Science 2777: pp. 1260- 1261.

Hansen A. J. and Defries R., 2007. Ecological Mechanisms Linking Protected Areas to Surrounding. Ecological Application 17 (4): pp. 974- 988.

Hardin G., 1968. The Tragedy of the Commons. Science 162: pp. 1243- 1248.

Hartig F. and Drechsler M., 2009. "Smart Spatial Incentives for Market-based Conservation". Biology of Conservation 142 (2): pp. 779- 788.

Houck O., 2002. Aftershock and Prelude: The Clean Water Act TMDL Program V. Environmental Law Reporter 32: pp. 10385- 10419.

Huber R. M., Ruitenbeek J., and Seroa da Motta R., 1998. Market-based Instruments for Environmental Policymaking in Latin America and Caribbean: Lessons from eleven countries. World Bank discussion paper No. 381, World Bank, Washington (D.C.).

IPCC, 2003. Good Practice Guidance for Land-use, Land-use Changes and Forestry. International Panel on Climate Change (IPCC), Vienna, Austria.

Jenkins M., Scherr S. and Inbar M., 2004. Market for Biodiversity Services. Environment 46(6): pp. 32- 42.

Jenssen M. and Hofmann G., 2004. First Results of the Environmental Monitoring of Forest Ecosystems in the Biosphere Reserve Spreewald. Beitrag für Forstwirtschaft und Landschaftsökologie, Jahresinhaltsverzeichnis 2004 (Heft 4): pp. 201.

Jentsch H. and Klaeber W., 1992. Neumanns Landschaftsführer Spreewald. Radebeul.

Johnson N., White A. and Perrot-Maitre D., 2001. Developing Markets for Water Services from Forests: Issues and lessons for innovators. Forest Trends, Joint published by World Resources Institutes and the Katoomba Group, Washington (D.C.).

Kleijn D. and Sutherland W. J., 2003. How Effective are European Agro-environment Schemes in Conserving and Protecting Biodiversity. Journal of Applied Ecology 40: pp. 947- 969.

Koji K., Shinsuku T. and Osamu O. (eds.), 2007. Systems Modeling and Simulation: Theory and application. Asian simulation conference 2006, Springer: pp. 103- 107.

Kotler P. and Armstrong G., 2001. Principles of Marketing. Ninth edition, Prentice-Hall, Inc.

Kotler P. and Nancy R. L., 2008. Social Marketing: Influencing behaviours for good. Third edition, SAGE publications.

Krausch H. D., 1960. Die Pflanzenwelt des Spreewaldes. Wittemberg Lutherstadt.

Krishnamurthy E. V., 1989. Parallel Processing: Principles and practice. Addison-Wesley Publishers Ltd.

Krishnam V. S., 2006. Study Guide for Use with Principle of Corporate Finance. The Mc Graw-Hill Companies.

Landell-Mills N. and Porras I., 2002. Market for Forest Environmental Services: Silver bullet or fool's gold? International Institute for Environment and Development (IIED), Earthscan, London.

Lauterbach S., 2007. An Assessment of Existing Demand for Carbon Sequestration Services. Journal of Sustainable Forestry 25 (1/2): pp. 75- 98.

Lindemann C., 1998. Performance Modelling with Deterministic and Stochastic Petri Nets. John Willey & Sons.

Ljungqvist L. and Sargent T. J., 2000. Recursive Macroeconomic Theory. Cambridge, Mass, MIT Press.

Löscher T., 2009. Optimisation of Scheduling Problems Based on Timed Petri Nets. ARGESIM Report, ARGESIM/ASIM.

Lowitzsch J., 2007. Financial Participation for a New Social Europe: A building block approach. "Financial Participation in the EU-27", The first international conference on financial participation in Berlin Germany, Conference edition, published by Inter-University Centre Split/Berlin.

LUA (Hrsg), 2004. Leitfaden zur Renaturierung von Feuchtgebieten in Brandenburg. Landesamt fuer Umwelt und Arbeitschutz (LUA), Studien und Tagungsberichte, Band 50: pp. 192.

Luce R. D., 1986. Uniques and Homogenity of Ordered Relational Structures. Journal of Mathematical Psychology 30: pp. 391- 415.

Mankiw N. G., 2008. Principles of Economics. Fifth edition, SOUTH WESTERN CENGAGE Learning.

Mclaughlin A. and Mineau P., 1995. The Impact of Agricultural Practices on Biodiversity. Agriculture, Ecosystems and Environment 55 (3): pp. 201- 212.

MEA, 2003. Human Welling-being: A framework for assessment. Millennium Ecosystem Assessment (MEA), Island Press, Washington (D.C.).

MEA, 2005. Ecosystem and Human Well-being: Synthesis. Millennium Ecosystem Assessment (MEA), Island Press, Washington (D.C.).

Milner R., 1989. Communication and Concurrency. Prentice-Hall.

MLUR, 2000. Rahmenempfehlungen zur Düngung 2000 im Land Brandenburg. Ministerium für Landwirtschaft, Umwelt und Ländliche Räume (MLUR), Landesamt für Ernährung und Landwirtschaft Frankfurt (Oder): pp. 60.

MLUR, 2003. Immissionsschutzbericht 2002: Potsdam. Ministerium für Landwirtschaft, Umwelt und Ländliche Räume (MLUR): pp. 88.

Myres N., 1996. Environmental Services of Biodiversity. Proceeding of the National Academy of Sciences of the United State of America 93 (7): pp. 2764- 2769.

Nair P. K. R., 1989. Agro-forestry Systems in the Tropics. Kluwer, Boston.

Natura, 2000. A European Network of Protected Sites. Established in compliance with EU-Directives 79/409/EEC (bird and habitat directives), European Union (EU).

Ngobo M., 2004. Natural Fallows of Southern Cameroon: Trends and implications for Agro-forestry Research. International Institute of Tropical Agriculture- Humid Forest Eco-regional Center, Yaounde, Cameroon.

OECD, 2001. Environmental Indicators for Agriculture: Methods and results. Volume 3, Organization of Economic Cooperation and Development (OCED), Paris.

OECD, 2007. Pollution Abatement and Control Expenditure in OECD Countries. Working Group on Environmental Information and Outlooks, ENV/EPOC/SE (2007)1, Organization for Economic Cooperation and Development (OECD), Paris.

Pagiola S. and Platais R., 2003. Payment for Ecosystem Services. World Bank, Washington (D.C.).

Paquin M. and Mayrand K., 2005. MEAs- based Markets for Ecosystem Services: Draft concept paper prepared for the Organization for Economic Cooperation and Development (OECD) workshop on Multilateral Environmental Agreements (MEAs) and private investment, Helsinki, Finland.

Pearl J., 2000. Causality: Models, Reasoning and Inference. Cambridge University Press.

Peterson J. L., 1981. Petri Net Theory and the Modeling of Systems. Prentice Hall, New York.

Petschick M., 2010. Biotoptypen Biosphärenreservat Spreewald. Landesamt für Umwelt, Gesundheit und Verbrauchtschutz, Brandenburg.

Petri C. A., 1962. Communication with Automata: PhD Thesis. University of Bonn, Germany.

Pipino L. L., Lee Y. W. and Wang R. Y., 2002. Data Quality Assessment. Communications of the ACM 45 (4): pp. 211- 218.

Ranganathan J. and Irwin F., 2007. Restoring Natural Capital: An action agenda to sustain ecosystem services. World Resourecs Institute, Washington (D.C.).

Ranganathan J., Munasinghe M. and Irwin F. (eds.), 2008. Policies for Sustainable Governance of Global Ecosystem Services. Edwarg Elgar Publishing Ltd, Cheletenham.

Rao P. K., 2000. Sustainable Development: Economic and policy. Black Well publishers Ltd, Oxford.

Reisig W., 1983. System Design using Petri Nets. Springer- Verlag, Berlin.

Reisig W. and Rozenberg G. (eds.), 1998. Lectures on Petri Nets I: Advances in Petri nets. Lecture Notes in Computer Science 1491.

Roemer J., 1996.Theories of Distributive Justice. Harvard University Press.

Rohr C., Marwan W. and Heiner M., 2010. Snoopy: A unifying Petri net framework to investigate bio-molecular networks. Bioinformatics 26 (7): pp. 974- 975.

Rubinstein A., 1998. Modeling bounded rationality. The MIT Press.

Sajoughian H. S. and Cellier F. E., 2000. Discrete Event Modelling and Simulation Technologies: A tapestry of systems and AI-based theories and methodologies. Springer- Verlag, New York, Inc.

Samuelson P., 1954. The Pure Theory of Public Expenditure. Review of Economics and Statistics 36: pp. 387- 389.

Shwartz M., 2000. Find a Market for Ecosystem Services: How much is an ecosystem worth? The New York City experience, Press release by Stanford University 29 Nov. 2000.

Sloane E. B. and Gelhot V., 2004. Application of the Petri Net to Simulate, Test and Validate the Performance and Safety of Complex, Heterogeneous, Multi-modality Patient monitoring Alarm Systems. Conference Proceeding, IEEE Eng. Med. Biol. Soc. 5: pp. 3492- 3495.

Slokoe J. and Teague E., 1995. Integrated Solid Waste Management for Rural Area: A planning trial kit for solid waste managers. Land- of- sky Regional council 25, North Carolina, Heritage Drive Asheville 28806 (704): pp. 251- 6622.

Slootweg R. and van Beukering P., 2008. Valuation of Ecosystem Services and Strategic Environmental Assessment: Lessons from influential cases. Netherlands Commission for Environmental Assessment, Utrecht.

Smith P. and Begg D., 2000. Economics. Sixth edition, McGraw-Hill, International.

Stephens J., 2006. Growing Interest in Carbon Capture and Storage (CCS) for Climate Change Mitigation: Sustainability. Science, Practice and Policy 2(2): pp. 4- 13.

Stiglitz J., 2000. Economics of the Public Sector. Third edition, Norton Press.

Succow M., 1992. Biosphärenreservat Spreewald: Labyrinth der tausend fließe. In Succow M. (Hrsg), München (Tomus verlag), Unbekanntes Deutschland: pp. 132- 151.

Ten-Brink B., 2000. Biodiversity Indicators for the OCED Environmental Outlook and Strategy: A feasibility study. National Institute of Public Health and the Environment, the Netherlands, RIVM Report 402001014.

Tietenberg T., 2002. The Tradable Permits Approach to Protecting the Commons: What have we learned? Nota di Lavoro 36.2002, Fondazione Eni Enrico Mattei, Venice (Italy).

UNCED, 1992. World Earth Summit Agenda 21: Program of action for sustainable development. United Nation Conference on Environment and Development (UNCED), Rio de Janeiro 3- 4 June 1992, United Nations publication.

UNECE, 2006. Forest products annual market review 2005-06. Geneva Timber and Forest Study paper 21, United Nations Economic Commission for Europe (UNECE) /Food and Agricultural Organization (FAO), New York and Geneva.

UNEP, 2005. Environment for Development: Pro- poor market for ecosystem services. Published by the United Nations Environmental Program (UNEP).

Vahrson W. G., Luthardt V. and Dreger F., 2000. Flächenauswahl und Ökosystem Monitoring in den Biosphärenreservaten Schorfheide- Chorin und Spreewald. Umweltwissenschaften und Schadstoff- Forschung 12 (6): pp. 362- 372.

Van Bueren M., 2001. Emerging Markets for Environmental Services: Implications and opportunities for resource management in Australia. Publication no. 01/162, Rural Industries Research and Development Corporation (RIRDC), Barton, Australia.

Vedder A. and Wright M., 2003. Mapping the Conservation Landscape. Conservation Biology 17 (1): pp. 116- 131.

Vitousek P., 1997. Human Domination of Earth's Ecosystems. Science 277: pp. 494- 499.

Von Eye A., 2003. Review of Cliff and Keats, Ordinal Measurement in the Behavioral Sciences. Applied Psychological Measurement 29: pp. 401- 403.

von Weizsächer E. U., Young O. R. and Finger M., 2005. Limits to Privatization: How to avoid too much of a good thing. Earthscan, London.

Wätzold F., 2006. Organizing a Public Ecosystem Service Economy for Sustaining Biodiversity. Ecological Economics 59 (3): pp. 296- 304.

WCED, 1987. Our Common Future: The Brundtland Report. World Commission on Environment and Development (WCED).

WEO, 2008. International Energy Agency Publication: World Energy Outlook (WEO), Organization of Economic Cooperation and Development (OCED), Paris.

Wilkinson J. and Kennedy C., 2002. Banks and Fees: The status of off-site wetland mitigation in the United States of America (USA). Environmental Law Institute, Washington (D.C.).

Willer H. and Yussefi M., 2006. The World of Organic Agriculture: Statistics and emerging trends 2006. International Federation of Organic Agriculture Movements (IFOAM), Bonn (Germany) and Research Institute of Organic Agriculture, Frick (Switzerland).

Wil V., José-Manuel C., Fabrice K., Gabriela k. and Daniel M., 2003. Petri Net Approaches for Modelling and Validation. LINCOM Studies in Computer Science 01, LINCOM Europa.

Winkler W. E., 2004. Method for Evaluating and Creating Data Quality. Information Systems 29 (7).

Wissel S. and Wätzold F., 2008. Applying Trade Permits to Biodiversity Conservation: A conceptual analysis of trading rules. Enviromental Research Centre Discussion Paper, 7/2008, published by Enviromental Research Centre, Leipzig.

Wohlmuth K., 2007. Africa- Commodity Development, Resource Curse and Export Diversification: African Development Perspective Year book 2007. Volume 12, LIT-Verlag, Münster.

World Agro-forestry Center, 2007. Medium-Term Plan 2008-2010: Transforming lives and landscapes through agro-forestry sciences. Published by World Agro-forestry Center, Nairobi (Kenya).

World Resources Institute, 2008. Roots of Resilience: Growing the wealth of the Poor. Published by the World Resource Institute, Washington (D.C.).

Wunder S., 2005. Payment for Environmental Services: Some nuts and bolts. CIFOR discussion paper, Center for International Forest Research, Jakarta, Indonesia.

Zeigler B. P., 1976. Theory of Modelling and Simulation. John Wiley & Sons, Inc.

Zoltan J. A. and Gerlowski D. A., 1996. Managerial Economics and Organization. Upper Saddle River, New Jersey. University of Baltimore Prentice Hall, Prentice- Hall, Inc.

Internet Sites:

Bio- trade, 2008. Website Bio-trade, http://www.biotrade.org (accessed 21 Dec. 2009).

BSDglobal, 2007. Business and Sustainable Development. http://www.bsdglobal.com (accessed 28 Nov. 2007).

Burkhard B., Kroll F., Müller F. and Windhorst W., 2009. Landscapes' Capacities to Provide Ecosystem Services: A concept for land cover based assessments. Landscape Online, The official Journal of the International Association for Landscape Ecology, Chapter Germany (IALE-D), http://www.landscapeonline.de/ (accessed 30 May 2010).

Conservational International, 2007. World's First Publication on Sustainable Hotel, Sitting, Design and Construction Guiding Principles. http://www.celb.org/xp/CELB/news-events/press_releases/09142005.xml (accessed 21 Sept. 2008).

Convention on Biological Diversity, 2004. Guidelines on Biodiversity and Tourism Development. http://www.biodiv.org/doc/publications/tou-gdl-en.pdf (accessed 15 June 2005).

Eco-trade, 2009. EcoTRADE: Market- based instruments for cost- effective biodiversity conservation. http://www.ecotrade.ufz.de (accessed 18 Dec. 2009).

EIA, 2008. World Crude Oil Prices. Energy Information Administration (EIA) Webpage, http://tonto.eia.doe.gov/dnav/pet/pet_pri_top.asp (Release Date: 16 Oct. 2008 and accessed 17 Oct. 2008).

ESRL, 2008. Trends in Atmospheric Carbon Dioxide. Mauna Loa, Earth System Research Laboratory (ESRL) Webpage, http://www.esrl.noaa.gov/gmd/ccgg/trends/index.html (accessed 20 Oct. 2008).

European Science Foundation, 2007. Buying and Selling Habitants to Help Wildlife. http://www.sciencedaily.com/releases/2007/10/071012113048.htm (Science Daily retrieved 05 October 2009 and accessed 20 Feb. 2010).

IAB, 2005. Webpage of Inter-American Bank (IAB) http://www.iab.org/sds/ENV/site_45_e.htm (accessed 20 Sept. 2006).

IFC, 2006. Biodiversity Conservation and Sustainable Natural Resource Management. Performance Standard 6, International Finance Corporation (IFC), Washington (D.C.), http://www.ifc.org/ifcext/enviro.nsf/AttachmentsByTitle/pol_PerformanceStandards2006_PS6/$FILE/PS_6_BiodivConservation.pdf (accessed 20 Dec. 2006).

IFC, 2008a. Website International Finance Corporation (IFC), http://www.ifc/ifcext/enviro.nsf/Content/CarbonFinance (accessed 20 Sept. 2 008).

IFC, 2008b. Environment: Carbon finance. Intra- International Finance Corporation (IFC), http://carbonfinance.org/working paper for IFC (accessed 20 Sept. 2008).

MEAs, 2005. Multilateral Environmental Agreements and Private Investment: Business contribution to address global environmental problems. Multilateral Environmental Agreements (MEAs), P.61, ENV/EPOG/GSP (2004)4 Final, http://www.oecd.org/dataoced/46/45/34860486.pdf (accessed 11 Sept. 2008).

Munichre, 2000. Munichre Press Release. http://www.munichre.com (accessed 20 July 2008).

Steiner R., 2005. 1% Earth Profits Fund: A private- sector conservation finance initiative for the 21st century. International Union for Conservation of Nature (IUCN), Commission on Environmental, Economic and Social Policy, Gland (Switzerland), http://www.iucn.org/themes/ceesp/Wkg_grp/Seaprise/Ref%205%20Earth%20Profits%20Fund.doc (accessed 29 Jan. 2006).

ten Kate K., Bishop J. and Bayon R., 2004. Biodiversity Offsets: Views, experience and the business case. Insight Investment, International Union for Conservation of Nature (IUCN) Gland (Switzerland) and Cambridge and London (UK), http://www.eldis.org/static/DOC16610.htm (accessed 30 Dec.2005).

UNCTAD, 2009. Website: http://www.unctad.org/ghg/index.htnl (accessed 10 July 2009).

UNFCCC, 2005. Afforestation and Reforestation CDM Project Activities. United Nations Framework Convention on Climate Change http://cdm.unfccc.int/methodologies/ARmethodologies/projects/pac/pac_ar.html (accessed 25 Apr. 2005).

World Bank, 2008. Carbon Finance for Sustainable Development. Carbon Finance at the World Bank, http:// www.carbinfinance.org (accessed 16 Sept. 2009).

World Ecotourism Summit, 2002. Webpage of World Tourism Organisation. http://www.world-tourism.org/sustainable/IYE/quebec/anglais/index_a.html (accessed 16 Dec. 2009).

http://www.IFOAM.org (accessed 20 Feb. 2010).

http://www.world-tourism.org (accessed 23 Feb. 2010).

http://www.rainforest-alliance.org/tourism.cfm?id=main (accessed 05 Feb. 2008).

http://www.spreewald-erlebnis.de (accessed 13 May 2010).

http://www.un.org/esa/dsd/agenda21/ (accessed 20 Jan. 2009).

http://biodiversity-chm.eea.europa.eu/information/document/F1088156525/F1125582140 (access 11 Aug. 2010).

http://www.citypopulation.de/php/germany-brandenburg.php (access 02 October 2010).

http://geoportal.lkspn.de (access 03 Oct. 2010).

i want morebooks!

Buy your books fast and straightforward online - at one of world's fastest growing online book stores! Environmentally sound due to Print-on-Demand technologies.

Buy your books online at
www.get-morebooks.com

Kaufen Sie Ihre Bücher schnell und unkompliziert online – auf einer der am schnellsten wachsenden Buchhandelsplattformen weltweit! Dank Print-On-Demand umwelt- und ressourcenschonend produziert.

Bücher schneller online kaufen
www.morebooks.de

VDM Verlagsservicegesellschaft mbH
Heinrich-Böcking-Str. 6-8　　Telefon: +49 681 3720 174　　info@vdm-vsg.de
D - 66121 Saarbrücken　　　Telefax: +49 681 3720 1749　　www.vdm-vsg.de

Printed by Books on Demand GmbH, Norderstedt / Germany